Patron-Client State Relationships

Patron-Client State Relationships

Multilateral Crises in the Nuclear Age

By

Christopher C. Shoemaker
and
John Spanier

PRAEGER

PRAEGER SPECIAL STUDIES • PRAEGER SCIENTIFIC

New York • Philadelphia • Eastbourne, UK
Toronto • Hong Kong • Tokyo • Sydney

PRAEGER—SHOEMAKER: Patron-client state relationships—Copyright

Library of Congress Cataloging in Publication Data

Shoemaker, Christopher C.
Patron-client state relationships.

Bibliography: p.
Includes index.
1. International relations. 2. Balance of power. 3. Atomic weapons. 4. United States–Foreign relations–Soviet Union. 5. Soviet Union–Foreign relations–United States.
I. Spanier, John W. II. Title.
JX1391.S554 1984 327.1'1 83-17822
ISBN 0-03-063881-X

Published in 1984 by Praeger Publishers
CBS Educational and Professional Publishing
a Division of CBS Inc.
521 Fifth Avenue, New York, NY 10175 USA

3456789 052 987654321

Printed in the United States of America
on acid-free paper

CONTENTS

Patron-Client State Relationships

CHAPTER 1

INTRODUCTION: WHY ANOTHER CRISIS STUDY?

BIPOLARITY AND NUCLEAR ARMS

Two of the principal features that shaped the character of the cold war were the bipolar balance of power and the growing stockpiles of nuclear weapons. Together, these two factors made the period after World War II extremely dangerous, creating conditions in which global war was both probable and potentially suicidal.

Why was war such a distinct possibility? The overriding reason is that a bipolar distribution of power is more sensitive than any other distribution to vagaries in the international environment. Bipolarity has generally been considered to be the most unstable distribution of power or, more succinctly, the most likely to erupt in war. In a two-power relationship, each state is continuously concerned with the balance of power, the prerequisite for its security. Each fears not only that the balance may be upset—a gain of power and security for one will mean a loss of power and security for the other—but that such a loss may be irreversible. In a biploar distribution, in which each of the two powers relies primarily upon itself for its strength, there is no third power to throw its weight into the scales and restore the balance. In these circumstances, each power is fearful of any shift in the balance; even relatively small shifts are not generally allowed lest the loss of one "domino" lead to the loss of further dominoes, resulting cumulatively in a major loss of power. Thus when one power pushes, the other must push back. Bipolarity, to sum up, means constant confrontation and frequent clashes as one side challenges the status quo and the other responds. Crises, even limited wars, are the result. A bipolar distribution not only permits such conflict, it encourages it. Indeed, hanging over this bipolarity is the mutual attractiveness of a preventive war. When each power sees the other as the only real threat to its security, the temptation to strike preemptively and abolish that threat is high.

Nuclear weapons, superimposed upon such a bipolar rivalry, mean that the potential consequences of this two-power competition are awesome. Not only does bipolarity intensify competition and increase the chances of conflict, the existence of massive stockpiles of nuclear weapons makes that competition all the more risky. Yet, while the fear of a first strike by the opponent and the possible need "to get there firstest with the mostest," and the likelihood of intermittent crises and occasional limited wars—all of which could erupt or escalate into nuclear warfare—are ever present, the devastating nature of nuclear war has also led the superpowers to discipline their use of power. The emphasis on *deterring* nuclear war, on *crisis management* and *limited (conventional) war* testifies to this shift away from the use of force to the use of the threat of force or coercion; and when force is used, the emphasis is on the limitations on this usage and the need to avoid or control escalation. In these circumstances, an increasingly widely accepted thesis has focused on the disutility of the use of force. Whether this is a valid thesis or not, it does seem that while bipolarity is unstable and more likely to result in war than other distributions of power, *nuclear* bipolarity appears to be remarkably stable, if a chief criterion of judgment is avoiding the eruption of another total war. Although the causal relationship between stability and nuclear bipolarity is intuitive, the fact remains that the superpowers have avoided a major war for four decades.

SUPERPOWER CRISIS MANAGEMENT

One reason for the apparent stability of nuclear bipolarity is that not only have the United States and the Soviet Union been able to deter one another from using nuclear weapons, but they have been able to resolve their bilateral crises peacefully. Such crises have occurred when one power has taken an action that the other has seen as a threat to its vital interests; it, therefore, reacted in order to preserve or restore the status quo.[1] A perceived threat to vital interests is one characteristic of a crisis and, along with surprise (the adversary's move was not expected at the time) and the policymakers' perceptions that they must act quickly, leads to the most important feature of a crisis: the heightened expectation of violence. This expectation, of course, underlies much of traditional international politics; in a crisis, however, this expectation rises to prominence. A crisis therefore occurs in the transition zone between peace and war.

Indeed, it has been suggested that in the nuclear era, crisis management is a substitute for war. In prenuclear times, when one state was determined to change the status quo, and the challenged state refused to comply, war would probably result. But since nuclear war is no longer considered a rational instrument of policy, crises become the means for challenging or preserving the status quo. In such confrontations, each power tests the other by taking the measure of its opponent (including its firmness in attaining or defending its interests and its willingness to take risks) in what Schelling has called the "manipulation of risk."[2] Successes or setbacks on the individual moves either side makes are equivalent to victories or defeats in war; each move becomes a test. In brief, when it is too dangerous to fight, political maneuver becomes a substitute for military battles. The key element in all this becomes each state's perception of the other's resolve to win, or not to lose.

Bilateral nuclear crises, once they erupt, are inherently very difficult to resolve but, at the same time, tend to have built-in dynamics for escalation control. On the one hand, one power wants to prevail and compel the adversary to give in to its demands; it wants to "win" and expand its influence. On the other hand, neither side wants nuclear war. Thus they are both rivals and partners simultaneously; rivals as one seeks to gain at the expense of the other, partners in cooperating to avoid nuclear war. In the former role, they seek to make their positions credible by threatening to resort to force; in the latter role, they seek to find ways of resolving their dispute without having to use force. Crisis management is the effort "to balance and reconcile these diverse elements" of bilateral competition and the shared danger of war.[3]

Studies of crisis management, therefore, focus on tactics.[4] While not neglecting the important contexts within which crises occur, these studies necessarily concentrate on the "bargaining process": the role of initiative, demonstrating both resolve and prudence simultaneously, how to communicate or signal effectively, maintaining one's freedom of choice and avoiding becoming trapped in any position, ways to increase the credibility of one's commitments and threats, and various other tactics of "coercive bargaining." In the prenuclear age, nations locked in crises were more likely to resort to force quickly; the penalty for doing so was not suicidal. But after World War II, as nuclear war hung over superpower crises, the superpowers, reluctant to resort to force if avoidable, raised the threshold at which force would be used. In this way, they increased the range of coercive moves short of war with which they can test one another in what Glenn Snyder refers to as the substitution of "psychological force for physical force".[5] Implicit in these analyses of tactics is the bipolar nature of the crisis—that is, there are only two sets of decision makers who can concentrate on each other. This at least reduces the chances of misperception and miscalculation as well as reducing the number of variables impinging on the crisis.

BIPOLYCENTRIC CRISES

While bipolar crisis management has been widely studied, based obviously on a fair number of actual cases, there are equally obviously few studies of the type of crises more likely in the future: bipolycentric crises. Instead of direct bipolar American-Soviet confrontations, bipolycentric crises start with political and military conflicts among regional states that should not, on their own merits, draw the superpowers into direct conflict. However, the competitive nature of their relationship has tended to do so. The world has changed significantly since the analytically elegant and classic bipolar Cuban Missile Crisis, and these changes have drastically altered the substance and venue of superpower crises. These changes include: (1) the birth of a large number of new non-Western states; (2) the multiplication of regional rivalries among the indigenous states for power, prestige, primacy, and security; (3) the changing strategic balance between the United States and the Soviet Union; and (4) the emergence of the Soviet Union as a global power.

The first two changes require little comment. The multiplication of states, the rise of national assertiveness, the clashes of competing "national interests" are evident in all regions in the Third World, especially the Middle East, Southwest and Southeast Asia. The last two changes need further elaboration, however, for it is they that threaten to turn essentially regional quarrels, motivated by local concerns, into global conflicts involving both superpowers. It is in this respect that the shifting super power military balance becomes critical. The strategic balance has changed from one of U.S. nuclear superiority to superpower parity. Many observers fear that the trend is toward Soviet strategic superiority, and some, including President Reagan, argue that the Soviet Union has already attained this ascendancy. For our present purposes, we need merely note the loss of U.S. strategic superiority and the increasing instability of a deterrent balance in which the land-based intercontinental ballistic missiles (ICBM) forces of the United States—and of the Soviet Union—are perceived to be increasingly vulnerable to a first strike.

The chief consequence of this shift in the strategic balance, however, is as much psychological and political as military. Mutually recognized U.S. strategic superiority made the decisions of American policymakers to respond to Soviet challenges during the cold war period less agonizing because the risks and costs were lower than today; the steady growth of Soviet strategic power will now make Washington more hesitant about reacting to Soviet moves in the future. By the same token, one major reason why Soviet leaders in the past were cautious in their conduct of crises, if less cautious in precipitating them, had to be Soviet strategic inferiority. If the United States responded to a Soviet challenge, the Soviet Union had to find ways out of the crisis, either by retreat and some form of face-saving compromise (as in Cuba in 1962) or by allowing the crisis to simply fade away on the basis of the status quo (as happened with the several Berlin crises between 1958 and 1962). The Soviet attainment of at least parity has presumably enhanced the Kremlin's sense of military strength and confidence.

Indeed, the Soviet leaders see the emergence of strategic parity as a key development, to be ranked with the birth of the Soviet Union as the nucleus of a new postcapitalist socialist work order, the emergence after World War II of a socialist block, and the breaking of the capitalist encirclement of Soviet Russia as the single socialist state. The Soviet leaders have specifically claimed that the change in the strategic balance is responsible for three developments: (1) no important problems in the world can be settled without Soviet participation; (2) the United States during the 1970s become more restrained and did not escalate its activities or intervene in South Vietnam when it collapsed, in Angola and Ethiopia during the Soviet-Cuban intervention, in Iran when the Shah was overthrown, or in Afghanistan after the 1978 coup or the Soviet invasion a year later; (3) the Soviet Union and its friends can take more initiatives to expand their influence. For, while nuclear war is no more desirable now than before, the risks of Soviet and Soviet-sponsored interventions have decreased while those of the United States, if it reacts, have risen.

Thus, while some U.S. policymakers may in retrospect claim that the decisive element in resolving the Cuban Missile Crisis successfully was America's conventional superiority,[6] the steady Soviet buildup since the crisis and subsequent claims about beneficial results of its strategic growth certainly suggest that the Soviets think otherwise. They attribute the United States demand that they withdraw their missiles from Cuba, and Soviet compliance to American strategic superiority. The same is true in other areas of confrontation. After all, during all the years they sought to evict Western forces from West Berlin, did they not possess overwhelming local conventional superiority?

The fact, however, is that today the Soviet Union not only basically neutralizes U.S. strategic power, but it has also built up and modernized its conventional forces. In 1962 the Soviet Union could not support its Cuban operation with its navy. The Soviet military growth since then has, therefore, included building a sizable modern surface navy and airlift capability. Historically, Russia, Tsarist and Soviet, has been a land power that, in its expansionist phases, has pushed outward around its borders. Essentially, Russia has been a Eurasian power. The newly acquired Soviet conventional capability is transforming the Soviet Union from a regional land power to a global land/sea power. The capacity to project Soviet influence beyond Eurasia is likely to expand in the future; the Soviet armaments program has shown few signs of slowing the momentum it has established over the past two decades.

Thus the emerging polycentric world—composed of many countries, most of them relatively small but capable of deciding to go to war with neighbors—poses enormous dangers. For the United States and the Soviet Union, both now global powers, have corresponding interests. Few conflicts in distant areas of the world beyond Europe do not inpinge on their interests; some of them will surely affect vital interests according to their leaders' calculations and perceptions. Such conflicts are thus likely to draw in the superpowers, as in 1914 the local quarrel between Serbia and Austria-Hungary on the fringe of the European system drew in all the great powers, precipitating a four-year slaughter unparalleled in Europe since the thirty-year religious wars of the early seventeenth century.

Managing such polycentric crises by the superpowers will be significantly more difficult than resolving bipolar crises. The latter involves only two actors, who need only watch each other and who, over a series of crises, have learned a lot about each other's crisis behavior and management; they also were aware of their stake in a peaceful outcome. Current and future regional conflicts are likely to be the product of states absorbed in their quarrels with one another, much less concerned with the impact of their differences on global peace and survival. These local states will seek to attract superpower assistance and involvement to win or at least to avoid losing; longer-run or broader global interests are understandably of less concern to them than are their immediate problems. They are local powers with local interests at issue that to them are vital but that have only marginal inherent impact on global affairs.

Beyond the priority of these concerns, the relationships involved are both greater in number and more complex: local state to local state, local state (or client as we will call it in this study) to each superpower (or patron), and patron to patron. In terms of this greater number of relationships than in a bipolar crisis, it should be obvious that the number of opportunities for misperceptions, miscalculations, and miscommunications rise correspondingly. Not only will the ability of the superpowers to control the outbreak of crises be relatively low (because they cannot control the actions of most other states), but their capacity for successful crisis management will also decline. A bipolycentric system is clearly more dangerous than a bipolar one.

The key factor that determines the influence flow is the nature of the patron-client relationship; under the most trying conditions, despite its greater power and resources, the patron may not be able to restrain its client. Indeed, the latter may actually be able to manipulate its patron more effectively than vice versa. This changing and reciprocal relationship is at the heart of the problem of future crisis management. While brilliant tactical improvisation, should such a bipolycentric crisis occur, may contribute to a resolution that protects specific client and patron interests, as well as the peace, this is not enough. As with the many studies of deterrence, arms control, crisis management, and limited wars, we need first to understand the concepts: the nature of patron-client relationships, the different types of such relationships that are possible, and then empirically study past crises that approximate the kind of bipolycentric crises that may occur in the future. It is in this spirit that we undertook this study: to point to a critical problem and, hopefully, advance our understanding of the issues involved so that the United States and the world can be spared the horror of a nuclear war in a world in which smaller powers pursue their interests with as much dedication as the great powers.

NOTES

1. See Glenn H. Snyder and Paul Diesing, *Conflict Among Nations* (Princeton: Princeton University Press, 1977), pp. 1–32, for an anatomy of international crises.

2. Thomas C. Schelling, *Strategy of Conflict* (Cambridge, Mass.: Harvard University Press, 1960).

3. Phil Williams, *Crisis Management* (New York: Wiley, 1976), p. 29.

4. Snyder and Diesing, *Conflict Among Nations*; Oran R. Young, *The Politics of Force* (Princeton: Princeton University Press, 1968); and Coral Bell, *The Conventions of Crisis* (New York: Oxford University Press, 1971).

5. Snyder and Diesing, *Conflict Among Nations*, p. 459.

6. Essay: "The Lessons of the Cuban Missile Crisis," *Time*, September 27, 1982, pp. 85–86 (written by Dean Rusk, Robert McNamara, and McGeorge Bundy, President Kennedy's secretaries of state and defense, and special assistant to the president for national security affairs, as well as other high Kennedy officials).

CHAPTER 2

THE RELATIONSHIP: A CONCEPTUAL SKETCH

The first step in examining patron-client state relationships and their impact on global stability is to understand the foundations of such relationships and the dynamics by which they change. Patron-client state relationships are a reality in the present global political system, and the dangers of bipolycentric crises have been repeatedly in evidence.

For example, within the span of but a few weeks in October 1973, the United States and the Soviet Union transformed their bilateral relationship—temporarily only, admittedly—from détente to a military confrontation accompanied by a heated verbal exchange reminiscent of the most serious bipolar crises that marked the cold war. This was the result not of any direct, deliberate move initiated by one superpower against the other. Rather, this was the product of the war between Israel and the Arab states, a conflict that neither superpower had actively sought. Yet, despite the almost total dependence of all the involved belligerents on the superpowers for weapons, spares, and replacements, Israel, Egypt, and the other Arab states pursued their own political courses of action that led to the conflict and were only marginally responsive to superpower demands during the war. The impact of the October War spread far beyond the confines of the Middle East and had potentially cataclysmic implications for the entire world.

How, then, are the actions of small Third World states able to involve and influence the superpowers who, by right of their dominance of the international arms markets and their awesome military power, should be able to remain uninvolved and indeed control such actions? Why do the superpowers, with so much to lose, allow themselves to be subjected to any measure of control by Third World states? This book is intended to provide some answers to these questions through an examination of the genesis and dynamics of patron-client state relationships and a detailed application of the theoretical constructs to the relationship between the Soviet Union and Egypt during the period 1967 to 1973.

THE INTERNATIONAL MILIEU

Critical to the understanding of patron-client state relationships is an examination of the global political and strategic environment of which such relationships are integral parts. Although particular patron-client state relationships may be treated in isolation for their historical interest, an analytical understanding of such ties is wholly dependent upon an understanding of their links to the broader strategic balance.

The strategic parity situation discussed above has translated itself into relative stability in Central Europe, an area that each superpower has perceived as an area of highest interests to its security.[1] Instead of mitigating or eliminating the competition between the United States and the Soviet Union, however, the strategic nuclear balance has shifted the superpower competition to other regions of the world, such as the Middle East, Southern Africa, the African Horn, Southwest Asia, and the Persian Gulf. The relationships between the superpowers and the smaller states of these areas are thus defined by the context of the superpower competition. It seems doubtful, for example, that the Soviet Union would have much interest in the Peoples Democratic Republic of Yemen (PDRY) were it not for the threat the PDRY poses to pro-Western Saudi Arabia or the potential for the PDRY to provide the Soviet Union with bases for interdicting the Western sea lines of communication (SLOCs) through the Bab-el-Mandeb and the Arabian Sea.

Smaller states, then, gain a value to each superpower based not only upon their intrinsic worth but also upon their ability to confer a competitive advantage on one of the superpowers. These smaller states become scarce resources that are available to the highest superpower bidder, and are sometimes able to extract a considerable price for their particular contribution to the superpower competition. Throughout any discussion of patron-client state relationships, then, it is critical to view seemingly irrational and irresponsible superpower interest in, and concessions to, a smaller state in the context of this acute superpower bilateral competition.

The result is a paradox. On the one hand, the purpose of close associations between the superpowers and smaller powers is to permit the superpower competition to occur in a relatively low-risk environment. On the other hand, as this competition is transferred to the Third World, the superpowers may become involved in the very sorts of bilateral confrontations each sought to avoid. When smaller Third World states seek the support of the superpowers, the latter—driven by their own compelling competition—may be drawn into a cycle of escalation. Each superpower feels it cannot afford to have its smaller associate or client defeated, especially at the hands of the other superpower and its client. A change in venue for superpower rivalry does not necessarily alter its intensity.

Increasingly recognizing their ability to turn superpower competition to their own advantage, Third World states have actively pursued policies of nominal nonalignment. Given their general lack of military strength against the superpowers, these states have traded their political favors for superpower support; by becoming active and willing participants in superpower competition, they gained influence beyond their own capabilities.[2] And, in the nuclear era, there are an increasing number of assertive Third World states, some led by irresponsible Qaddafhi-like leaders and all pursuing their "national interest." Any degree of client control over superpower military options or resources therefore has serious implications for the entire international system.

Thus, a meaningful examination of patron-client state relationships must consider the nature of the international milieu in which such relationships fit. These relationships are, to repeat, the means by which the larger powers compete and are, therefore, inextricably linked to the intensity of the competition between the patron states.

PATRON-CLIENT STATE RELATIONSHIPS DEFINED

For analytical purposes, a patron-client state relationship may be distinguished from other forms of bilateral interaction by the dominance of several key elements. First, there must be a sizable difference between the military capabilities of the states involved. The client cannot, by itself, become a major military power in the international community; nor can it, by itself, guarantee its own security. This means that the principal "security transfers" between patron and client are unidirectional in nature, flowing from the patron to the client.

Second, and more importantly, the client plays a prominent role in patron competition. The more advantage the patron gains over its competitor through its association with its client, the more the patron will value the relationship, often in apparent contradiction to the material benefits that the patron derives. This aspect of patron-client states relationships is demonstrated by the patron trading specific funds and items of military equipment for concessions from the client that will readily translate into advantages over the patron's opponents. This is the dimension that provides the client with its primary means of influence over its patron and determines the extent of patron interest in the relationship.

Third, there is a critical perceptual dimension to patron-client state relationships. This is derived from consistent association between the two states for a recognizable, if sometimes only brief, period of time. This association can take place at a variety of levels, but it must be apparent to other observers in the international system that the patron and client are closely tied together.

The extent and strength of a patron-client relationship are not necessarily related to the extent and strength of other forms of bilateral interaction between patron and client. There are cases in which the patron derives direct economic return for its contribution to the client, as well as gaining political advantage over its competitors. This adds to the potential influence the client can exert. This type of relationship, such as the United States maintains with Saudi Arabia, complicates, but does not change, the basic utility of this definition.

Other, non-patron-client state relationships rest far more on mutually beneficial economic or security interchange with less apparent asymmetry in contributions. These sorts of non-patron-client relationships are evident in the U.S. relations with its NATO partners.

At the other end of the spectrum, some states are not factors in patron competition and are, therefore, not clients in the sense of this definition. Some states in areas such as West Africa or Latin America have simply not of sufficient interest to any patron to warrant the development of a relationship.

The litmus test for determining the existence and extent of a patron-client state relationship is an evaluation of the durability and nature of a relationship in the absence of patron competition. The relationship between the United States and Japan, for example, would surely flourish, if somewhat differently, even without superpower competition. Soviet relations with Ethiopia, on the other hand, would certainly be dramatically different, were it not for Soviet competition with the United States. The latter is clearly a patron-client relationship, while the former is not.

THE MILITARY DIMENSION: ARMS TRANSFERS AND SECURITY GUARANTEES

Given the present distribution of military power, patron-client state relationships are primarily aimed at enhancing their respective security. Although there are other aspects of such relationships, security transactions are the most evident and pervasive. The client principally seeks technology, sophistication, and numbers in the supply of weapons by the patron, as well as security guarantees that the patron can provide, often at no expense to the client. The patron seeks, in exchange, specific types of goals to further its efforts vis-à-vis its principal competitors. Because of this, arms transfers and security guarantees are the most efficacious indicators of the nature and extent of a patron-client state relationship. Through an examination of arms transfers and security guarantees over time, such as will be presented in later chapters, the strength and impact of specific patron-client relationships may be studied.

Arms transfers, in a world of rampant instability, are a powerful tool of influence for the patron. Not only do such transfers provide the patron with a valued *quid* for the *quo* the client can offer, but they also give the patron a long-term voice in the client's security. Weapons do not operate by themselves, nor do they operate forever. This fundamental, if somewhat obvious, point gives the patron two important sources of leverage. First, the client must obtain training from the patron in order to operate its weapons effectively. This provides the patron with the opportunity to station advisers in the client's country and to attempt to politicize the client's officer corps in military schools run by the patron. Second, since weapons require spare parts and replacements, the patron retains control, often considerable, over the client by controlling the spare parts pipeline. Patron-client state relationships do not normally concern themselves with economic or political development. Although economic development issues are important in global affairs, they do not carry the same weight in patron-client relationships as security issues do.

HAZARDS IN PATRON-CLIENT ANALYSIS

There are two major analytical weaknesses that are most common in the study and the practice of patron-client state relationships. First, most analysts and policymakers tend to ignore the underlying goals that motivate a superpower and a Third World client to enter into a relationship. There is an implicit view that patrons seek to maximize their influence everywhere at once. There is too little consideration of the international political context in which such relationships occur, and therefore, some analyses tend to be rather sterile. Even when goals are studied, there is an assumption that the goal structures of the two states remain constant over time. Under this formulation, a client that is useful to a patron at one point in time will remain useful forever.

This leads to a second and more serious deficiency. Patron-client state relationships are often thought of as static in nature; changes in such relationships are regarded as aberrations. This assumption of rigidity ignores the most important dimensions of patron-client state relationships, reducing their study to a series of isolated case histories. Above all, it also results in inadequate attention to the mechanisms by which patron-client state relationships change. These mechanisms are often of greater significance than the nature of the relationship itself at any given point in time, for mechanisms of change can have important implications for the international system. The role of crises, the most important instrument of change in patron-client state relationships, is similarly left inadequately explained.

Instead of rigid interaction between two states, patron-client state relationships are in reality fuzzy, fluid, fluctuating partnerships, subject to constant change and only becoming sharply defined in the context of a crisis. Admittedly, this view greatly complicates the analytical and policymaking process in dealing with patron-client state relationships; it forces a thorough examination of goals, perceptions, and leadership personalities. This is not to say, however, that rigorous, useful analysis of patron-client state relationships is impossible. It does say that such analyses are fraught with difficulty and are not amenable to easy, quick comprehension.

Moreover, patron-client state relationships are considerably more complex than the inter-patron competition they serve. Certain behavior patterns have come to be expected of patron states, particularly the superpowers. No such expectations are universally applicable to client states, which usually operate from a different world view, from a different hierarchy of objectives, and from a different historical perspective. Client states are not simply small patrons; treating them as such is an analytically weak and politically risky approach.

GOALS IN PATRON-CLIENT RELATIONSHIPS

Any examination of patron-client state relationships must logically begin with a study of the goals sought by both states, since these goals form the foundation upon which the entire relationship is built. When these are examined, it is apparent that there is an incompatibility of the most basic goals between the two states in the relationship. The patron, whatever its specific objectives in the relationship might be, seeks to exert some degree of control over the client. This control can take many forms, but in general, it implies the surrendering of some measure of the client's autonomy to the patron. The Third World client, on the other hand, is acutely sensitive to real or imagined "neocolonial" control and seeks to establish and maintain its independence in world affairs, sometimes in an apparently irrational manner.[3] In this environment, the client attempts to guard its autonomy from outside control, though such control lies at the very root of the patron's most basic objectives in the relationship.

Although this basic conflict between patron and client can be masked during the actual negotiation and conduct of the affairs of the relationship, its ubiquitous nature means that all patron-client state relationships rest upon a tenuous foundation and are inherently unstable. Nevertheless, despite this basic incompatibility, the patron and client enter into relationships because of specific objectives that, for the moment, transcend the underlying antagonism.

Patron Goals

The patron will seek to exert influence and control over the client by striving for specific goals of different types. The nature of the relationship is shaped by the contribution the patron believes the client can make toward these goals as well as the importance of the goals themselves. If the client can provide some valuable advantage for the patron over the patron's adversary, the patron will be willing to pay a much higher price in the relationship. There are several types of general objectives, in this regard, that the patron may seek.

First, there are ideological goals in which the patron attempts to remake the client in the patron's own image in order to display its system as superior to those of its competitors. These goals may be characterized by demands for changes in the client's political structure, by the introduction of new economic practices, by changes in social mores, or by direct control over the client's domestic or security policies. The Carter administration, for example, codified ideological goals in its foreign policy through the formal advocacy of human rights. When goals of this nature dominate the patron's objectives in the relationship, the patron will demand rigid adherance to its dictates and will tolerate few digressions.

Since the advantage the patron gains over its competitors from a relationship in which it seeks ideological goals is difficult for the patron to measure and is, in any case, indirect in nature, the patron will generally not value such a relationship as much as one in which other, more important, goals are at stake. The patron will, therefore, work less hard and invest fewer resources to maintain this kind of relationship and will be more exacting in its demands on the client state.

When ideological goals dominate the patron's hierarchy, it will seek to present the client to the outside world as a showplace of the patron's ideology and political system. Advantages of the patron's system of government will be displayed with great flourish so that the client will appear to be an attractive model for other Third World states to emulate. Internal client conformity not only serves as a model, but it also provides the patron with a means for controlling the client, should the situation in the international environment change dramatically and the client assume an expanded role in inter-patron competition. In addition, relationships with ideological goals allow the patron to articulate the moral tones of its foreign policy without seriously jeopardizing more highly valued objectives in other relationships by investing an excessive amount of scarce resources. If the client diverges from its ideological subservience, the patron's goals will be thwarted, and the relationship will not endure.

The second type of goal that may be sought by the patron is international solidarity. These goals are manifested by such things as voting cohesion in the United Nations, by the signing of international agreements, by visits between heads of state, by client pronouncements of international support for the patron, and by perceptual association of the client with the patron. The patron seeks such goals because it desires to present the impression to the world that the client is a member of its bloc, or at least, that the client is not a member of an opponent's bloc. These goals become especially important in cases in which the client was previously identified with the other bloc and can be shown to have switched loyalties.

When the patron has objectives of international solidarity, the client will be permitted more latitude in its domestic activities and policies as long as the appearance of international solidarity is maintained. For a relationship that is based upon goals of international solidarity will be more highly valued by the patron than will be one that is based on ideological goals. The patron will derive more obvious and tangible benefits from the relationship in terms of its competition with another patron. Because of this, the patron will be more responsive to the client's demands and more willing to surrender some measure of control over its own resources to the client in order to preserve the relationship. Thus, the client will be given more leeway to get into trouble, and the patron will be more ready to help the client out, should the client's activities exceed its capabilities.

The third type of goal sought by the patron is strategic advantage. In goals of this type, the patron seeks to control a vital piece of terrain owned by the client in order to obtain some military advantage over the patron's opponents. The patron may also seek to control a resource that is vitally important to its adversary. This type of goal is distinguished from the normal bilateral economic relationship by the fact that the patron really does not need the resource for its internal use; instead, it seeks the ability to deny the resource to the opponent. The patron may also seek to use the client as a surrogate in regional conflicts, to exploit the client's role as a staging area for revolution, or to use the client's territory to station patron armed forces to block the spread of an adversary's influence. In all of these measures, the client's role in direct patron competition is substantial.

Goals of strategic advantage are manifested by patron demands for bases on the client's soil, access to various client facilities, and cooperation between patron and client armed forces. When such goals are sought, the patron will permit the client a wide range of internal and international activities as long as these activities do not threaten the patron's strategic advantage. The patron values this type of relationship most because it confers a directly measurable and internationally obvious political and military advantage over its opponents and therefore contributes directly to the patron's security. When the patron possesses goals of strategic advantage, it will go to great lengths to preserve the relationship, even to the extent of allowing the client some measure of access to the patron's political and military resources. The patron will also be far more willing to use its own armed forces directly to ensure that the strategic advantage it derives from the relationship is not lost.

However, notwithstanding the oft-cited myths about each superpower's ability to project its military capabilities, the option to intervene is often neither available nor appropriate for use with a Third World client. Even superpowers, with a near-monoply on internationally significant military power, cannot project and sustain major formations at great distances from their homelands without enormous costs. Since coercive tactics may not be useful in such cases, cajoling tactics may be the only option the patron has. The patron, greatly valuing the relationship and the strategic advantage it offers, will often have to pay a steep price to maintain the ties. It is these relationships that present the greatest dangers to the international system.

Patron Goals and Patron Controls

The measure of patron control over the client will depend upon the goals the patron seeks in the relationship and the client's ability to fulfill these goals. The relationship between the goals of the patron and its degree of influence over the client may be summarized as shown in Figure 2.1.[4]

Figure 2.1

Strong Patron Control		Weak Patron Control
+	————————	–
Ideological Goals	International Solidarity	Strategic Advantage

It must be emphasized again that the type of goal sought by the patron can change over the duration of the relationship, and, with changes in goals, demands made upon the client and the measure of patron influence will change as well. Moreover, the patron may seek several goals simultaneously. The patron's valuation of the relationship will then depend upon the client's ability and willingness to meet all these goals, although goals of strategic advantage will remain dominant.[5]

Client Goals

Just as the patron's goals determine the demands it will place upon the client, the client's goals shape the degree of patron demands it will be willing to accommodate. Client goals, however, differ substantially from those of the patron, due primarily to the vast military power differential between patrons and their clients. To the patron, the relationship is principally a means to compete with other patrons in a hopefully low risk manner. To the client, however, the relationship may be its source of national survival. For, although Third World states have only a marginal capability to affect the international system militarily, they can certainly deal each other mortal blows. From the outline presented above, it is apparent that client states often live in what they perceive to be extremely hostile security environments, often hemmed in on several sides by enemies with regionally significant military capabilities. For this reason, client goals center around the nature of the threat that the client believes exists to its nation and government. The larger the threat to its existence and the more the client sees its salvation in the hands of the patron, the more likely will the client be to accept the relationship on the patron's terms.

There are other high-threat conditions, especially those of an internal nature, in which the prudent course of action for the client would be to disengage from the patron. Under some circumstances, the patron's influence can become a rallying point for nationalist sentiment with the client state. For the most part, however, the client state focuses its security attention on the external threat, a focus more amenable to patron assistance.

Therefore, from the client's perspective, the relationship will be shaped by its threat environment. At a lower level of threat, other goals such as economic development, regional leadership, and international prestige may well emerge in the course of the relationship. Under these circumstances, the client will be far more difficult for the patron to manage. A sense of urgency is often lacking, and the client has the luxury to shop around among various patrons for the support it needs. It has less incentive to surrender its internal autonomy, its international independence, or its territorial integrity to the patron. Depending upon the patron's goals, the client has a far greater chance to dominate the relationship if it has a low-threat environment.

Under conditions of high threat, however, the client will be much more willing to meet the patron's demands. Indicators of such a condition include the mobilization of the client's armed forces, the presence of hostile forces in close proximity to its borders, the sudden influx of large numbers of weapons to the client's enemies, or unusual domestic violence within the client state itself. The client will probably require military equipment, support, or even the presence of the patron's forces in order to overcome the threat it perceives. A desperate client in a high-threat environment may be willing to grant considerable concessions to the patron, which it would not extend under less threatening conditions. Under such high-threat situations, the patron will probably dominate the relationship, at least for the duration of the client's high-threat environment, as long as the client perceives the patron as a source of its salvation.

Client Goals and Client Controls

The relationship between the client's goals that are functions of its threat environment, and the client's control over the relationship may be summarized as shown in Figure 2.2

Figure 2.2

Strong Client Control + Low-Threat Environment	———————	Weak Client Control − High-Threat Environment

For the international system, the situations in which the client has strong control over the relationship are the most dangerous and call for the closest policy attention.[6]

A Patron-Client Typology

By combining the two continua, an array of different types of patron-client state relationships may be established. The value of developing such a typology lies in its ability to help explain, at a given moment in history, the flow of influence in a relationship. Further, the typology will be helpful in understanding the potential impact of a given relationship on the international system and will facilitate prediction of the direction the relationship will take in the future. To the extent that the

Figure 2.3

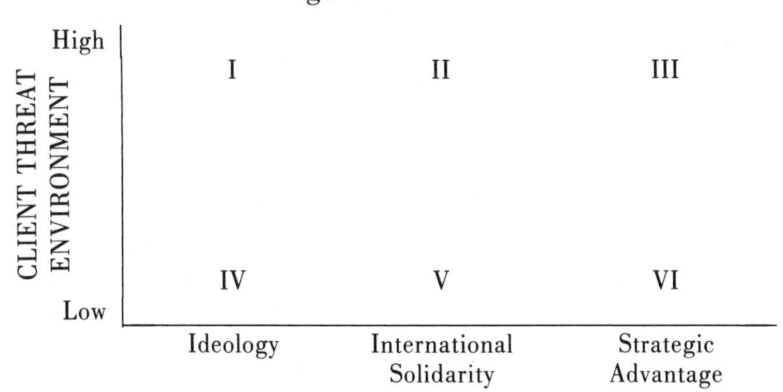

the typology accomplishes this purpose, it will be useful not only for heuristic purposes but also as a potential instrument for policymakers. As shown in Figure 2.3, this typology leaves us with six general types of patron-client state relationships, each with unique characteristics and implications for the international system.

CONCLUSION

Equipped with this typology, patron-client state relationships can be better understood as fundamentally bargaining relationships in which each state tries to extract from the other valuable concessions at a minimal cost.

The balance of this study is devoted to a detailed examination of the various types of relationships and the mechanisms by which they change. Illustrations will be presented to demonstrate the nature of patron-client state relationships and the flows of influence that determine the impacts a relationship may have on the international system.

In addition, the relationship between the Soviet Union and Egypt will be examined in some detail, with emphasis on the 1967 to 1973 period. This case study of a particular relationship shows the richness and complexity of patron and client interactions, as well as the dangers such relationships pose for the rest of the world. A full appreciation of the subtleties of patron-client state relationships is impossible without a firm grounding in historical experience. The case study reflects many of the theorectical dimensions of the nature of patron-client state relationships presented in succeeding chapters and, therefore, is a crucial adjunct to this effort.

The importance of patron-client state relationships in the international system of the future cannot be overstated. Scholars and policymakers need to devote far more attention to these relationships, both for the lessons they teach in international relations and for the major policy implications they have for the superpowers. In the absence of a thorough understanding of these relationships, unintended conflict and confrontation will continue to jeopardize stability and security throughout the world.

NOTES

1. The strategic balance is highly dynamic, and the present relatively equal distribution of nuclear capabilities is subject to rapid change. Stability in Europe is not necessarily guaranteed in an era of Soviet strategic nuclear superiority.

2. Robert L. Rothstein, *Alliance and Small Powers* (New York: Columbia University Press, 1968), p. 247.

3. Iran's self-emasculation as a result of its violent anti-American regime installed in January, 1979, is an example of this sort of apparent irrationality. U.S. policymakers, not fully understanding the depth of Iran's resentment of its client status, were caught repeatedly offguard by Iran's actions, including the seizure of the hostages.

4. It is apparent that the more the patron values the relationship, the weaker the patron's control will be. A useful mathematical analogy is $V_p = 1/C_p = (G_p) \times (F_c)$ where V_p is the value of the relationship to the patron, C_p is the patron's control of the relationship, G_p is the goal the patron seeks and F_c is the ability of the client to fulfill the goals sought by the patron. When G_p is large, such as when goals of strategic advantage are sought, even a small F_c will result in a high value for V_p and a concomitantly small C_p.

5. Our mathematical analogy in those circumstances becomes:

$$V_p = 1/C_p = (G_{p_1})\, X\, (F_{c_1}) + (G_{p_2})\, X\, (F_{c_2}) + (G_{p_3})\, X\, (F_{c_3})$$

where the numerical subscripts denote the several types of goals.

6. Again, the mathematical analogy may be useful. It goes like this:

$$V_c = 1/C_c = (T_c) \times (F_p),$$

where V_c is the value the client places on the relationship; C_c is the control the client can exert; T_c is the client's threat environment; and F_p is the client's perception of the patron's ability to meet the client's security needs.

CHAPTER 3

THE TYPOLOGY APPLIED

Patron-client state relationships are a pervasive element of the current international system; examples are evident in virtually every part of the Third World. Moreover, because of their role in superpower competition, they will continue to play a prominent role for the foreseeable future. It is, therefore, crucial to examine and understand such relationships in a systematic fashion. By examining examples of such relationships, using the typology outlined in the previous chapter, key characteristics may be discerned, characteristics that are useful for analytical, as well as policy, reasons.

RELATIONSHIP TYPES

Type I: Patron-centric

A type I relationship stems from patron goals of ideological conformity coupled with a high-threat environment for the client. The patron derives only marginal value from the relationship in its struggle against other patrons, while the client perceives its security as highly dependent upon patron support. The patron will control the relationship, exert the bulk of the influence, and extract significant ideological concessions from the client within a brief period of time. As long as the client perceives substantial threats to its security, and as long as the client perceives that the patron is able to alleviate the client's threat environment, a patron-centric relationship will endure.

Indications that a type I relationship has developed are: a rapid and massive governmental shakeup within the client country and the ascension to power of individuals agreeable to the patron; dramatic changes in the economic and social configuration of the client, including land reforms, nationalization of industry, and economic redistribution; major changes in long-standing client internal policies; and the release of the ubiquitous political prisoners, particularly those of the patron's political persuasion.

From the patron's standpoint, such a relationship would be marked by a modest increase in arms shipments and possibly advisory assistance, although probably not of the magnitude that the client desires. We would expect that even as the patron is addressing the client's security needs, the patron would not allow itself to become embroiled in a high-risk and dangerous confrontation with other patrons.

Relationships of this type are noted for their brevity. The client's concessions to the patron are the products of panic under fire. After the immediate threat has passed, the client will begin the process of undoing those levers of control that the patron has been allowed to establish.

The USSR and Ethiopia: July–December 1977

The Soviet-Ethiopian relationship during the period of July to December 1977 is an example of a type I interaction. Throughout these months, the Soviet Union was in a curious transition period in the Horn of Africa. It was attempting to support both Ethiopia and Somalia, ancient enemies that were locked in a major conflict in Ethiopia's Ogaden region. Moscow had no desire to sever its ties with Somalia, but it was exploring and expanding its relationship with Ethiopia.

During this period, the principal Soviet objectives in Ethiopia were ideological in nature. The Soviet-Somali naval facility at Berbera provided the support the Soviet Union needed for its Indian Ocean squadron, and Ethiopia had no comparable facilities. Moreover, the Soviet naval presence in the Indian Ocean at the time was still modest, easily supportable from Berbera. Therefore, the Soviet Union assessed the strategic value of Ethiopia as low.

The Soviet Union anticipated that it could maintain its access to Berbera and Mogadishu while supporting Ethiopia in the Ogaden. Moscow felt that Siad Barre had no other options than to accept Soviet support on Soviet terms.[1]

In Ethiopia, the Soviet Union saw an opportunity to establish a revolutionary government on the Soviet model and, in the process, demonstrate to other Third World leaders a blueprint for socialist revolution. It is significant to note that the early Soviet demands made of Ethiopia were not for bases, access, or military accords. Rather, Moscow pressed Ethiopia to create a socialist vanguard party that would penetrate into all levels of Ethiopian society and establish itself as the institution of the revolution.[2] This party would forge links to the Communist Party of the Soviet Union (CPSU) and would provide the Soviet Union with an additional lever of control over Ethiopia's future. Further, the Soviet Union saw a clear opportunity to demonstrate the dynamism of its appeal. Ethiopia, historically tied to the West and particularly to the United States, could be an example of a state that had thrown off the yoke of imperialism and embraced the socialist, nonaligned approach with great success.

On the other side of the relationship, Ethiopia found itself in an extremely high-threat environment. In the south and east, the Somali army, well equipped with Soviet weapons and supported by indigenous Somali guerrillas of the Western Somali Liberation Front (WSLF), had virtually ejected the Ethiopian army from the Ogaden by late summer 1977. In the north, the principal Eritrean secessionist groups had momentarily patched up their internal differences long enough to extend their control over large sections of Eritrea. Mengistu proved unable to deal effectively with either threat, and discontent with his leadership was growing within the Provisional Military Administrative Council (PMAC), Mengistu's primary instrument of power. As with many revolutionary governments, Ethiopia had sown the seeds for its own destruction through massive purges of all remnants of the old Haile Selassie regime, that resulted in massive military inefficiency and incompetence. The West, in particular the United States, had frozen arms shipments to Mengistu as his socialist rhetoric and anti-Western behavior alienated Ethiopia's erstwhile allies. In addition, the United States' strategic interests in Ethiopia had waned. Satellite technology had rendered the U.S. communications site at Kagnew far less important than it had been in the past. The United States was, therefore, less willing to tolerate Mengistu's behavior than it otherwise might have been.[3] Mengistu thrashed about in search of an arms supplier and a means to stem the Somali advance, and the Soviet Union stepped effectively into the equation with a substantial, although hardly overwhelming, infusion of weapons and supplies.

As might be expected in a type I relationship, the Soviet Union did not fully respond to Ethiopia's needs. Although Moscow stepped up its arms shipments, it did not engage in direct combat or allow its Cuban surrogates to do so until the Soviet-Ethiopian relationship underwent a fundamental change in November 1977. The catalytic event that brought about this change was the ejection of the Soviet Union from its facility at Berbera and the abrogation of the Soviet-Somali treaty of friendship and cooperation by Siad Barre.

The patron-centric description appears to fit this relationship over the period July-November 1977. The Soviet Union was able to begin molding and controlling internal Ethiopian affairs by responding to Mengistu's high-threat environment. It was able to extract concessions from Mengistu without ending its ties with Ethiopia's enemy, Somalia, and it was able to pay a relatively modest price directly to Mengistu for the internal concessions he extended. The Soviet Union, during the period, thoroughly dominated the relationship.

However, Soviet support for Ethiopia was not without its unintended, effects demonstrating the escalatory, uncontrollable nature of even the most patron-dominated relationship. By supporting Mengistu, the Soviet Union lost its naval facility at Berbera in which Moscow had invested considerable funds. It lost its chance to influence the United States to sign an Indian Ocean naval limitations treaty. During the massive air and sea lift to Ethiopia, the Soviet Union had nearly doubled its normal Indian Ocean naval strength, violating the spirit of mutual restraint that was the cornerstone of the negotiations. After this augmentation, the United States refused to resume the negotiations.[4] By the time Soviet activities resumed their precrisis levels, the United States had recognized the folly of the prospective treaty and declined to engage in further talks. The Soviet Union was also damaged in the Arab world. Even Iraq, hitherto a rigid supporter of the Soviet Union, quarreled with Moscow over its support of Christian Ethiopia over Muslim Somalia. Iraq became so incensed that it refused to grant the Soviet Union overflight rights for cargo aircraft directly destined for Addis Ababa.

All these costs clearly exceeded Soviet calculations, illustrating a fundamental point about patron-client state relationships: even in those cases in which the patron is clearly in command, the price of the relationship often escalates far beyond the patron's desires or control.

Type II: Patron-prevalence

In patron-prevalent relationships, the goals for the patron are client international solidarity. The patron will demand that the client adhere closely to the patron's international position, but it will be more willing to tolerate client independence in its own internal affairs. The patron will value the relationship more than it would a type I because of the greater advantage over the patron's opponents that can accrue.

As with a Patron-centric relationship, the client is faced with a high-threat environment. The client will make demands on the patron for direct security support, again usually in the form of weapons.

One of the clearest indicators of the emergence of a patron-prevalent relationship is a formal agreement or treaty between patron and client. The knowledge of the existence of such a document, even if its terms are not public, is a clear signal to the international community of the solidarity between patron and client.

Of particular note in this regard is the Soviet practice of signing treaties of friendship and cooperation with small, Third World clients. The Soviet Union signed ten such agreements during the 1970s, including treaties with Egypt (May 1971), India (August 1971), Iraq (April 1972), Somalia (July 1974), Angola (October 1976), Mozambique (March 1977), Vietnam (November 1978), Ethiopia (November 1978), Afghanistan (December 1978), and PDRY (People's Democratic Republic of Yemen) (September 1978).

In each of these cases, the existence of a treaty reflected diminished Soviet concern with internal ideological goals. In some cases, for example, in India, the Soviet Union ignored dramatic ideological differences. In other cases, such as in Afghanistan, Moscow actually discouraged ideological purity.

These treaties served two basic Soviet purposes. In some treaties, the primary thrust was to demonstrate to the international community the identification of the client with the Soviet Union. This was especially important when such treaties were made with former clients of the United States, such as Ethiopia.[5] Other treaties were more heavily skewed in the direction of providing specific benefits to the Soviet Union; these were normally in the form of secret access agreements. Regardless of the language of the treaty, however, the publicity with which the Soviet Union advertised the agreements spoke eloquently of Soviet desires to maximize perceptual association between Moscow and its client in the international community.

Both the United States and the Soviet Union place considerable emphasis on various other sources of international solidarity, such as pronouncements from various international organizations and the formal positions various states take on specific issues.[6]

Because the patrons in patron-prevalent relationships value the clients more, they are often more willing to respond to the client's perceived needs. The client in a high-threat environment will respond to patron demands with international solidarity. The influence flow is, then, less unidirectional than it was under a Patron-centric relationship. The patron is willing to give or support more to ensure the survival of the relationship.

USSR and Vietnam: 1978

The type II relationship between the Soviet Union and Vietnam was manifested in the Treaty of Friendship and Cooperation that the two states signed in November 1978. To the Soviet Union, Vietnam represented a very attractive area for demonstrating Soviet ascendancy over China in Southeast Asia. The treaty was a long-sought demonstration that Vietnam was firmly in the Soviet camp and that Chinese influence over the emerging Indochina regional power was minimal.[7] From the Soviet perspective, this gesture alone was worth the investment in arms for Vietnam and the risks of confrontation with China. There were other, more practical benefits that the Soviet Union eventually derived from the relationship, such as access to Vietnamese facilities at DaNang and Cam Ranh Bay, but, at the time the treaty was signed, these were only secondary objectives.

Closer examination of the treaty, especially as compared to other similar treaties, reveals the Soviet effort to publicly integrate Vietnam into its sphere of influence. There are more references to economic cooperation and political solidarity in the Vietnam treaty than appear in other treaties with Third World states. The formal inclusion of Vietnam in COMECON is, in this context, not surprising. The message in the treaty is clear. Vietnam recognized its special relationship with the Soviet Union and acknowledged its acceptance of Soviet, as opposed to Chinese, leadership in the Socialist world.[8]

For its part, Vietnam extracted invaluable support from the Soviet Union in meeting its high-security-threat environment caused by China. Prior to the end of the United States-Vietnam war, Vietnamese-Chinese hostilities remained latent, and Vietnam was able to extract political and military support from both the Soviet Union and China. At the end of the war, however, the greater U.S. threat was removed, and ancient regional animosities reasserted themselves. Kampuchea (Cambodia) under Pol Pot became a close ally of China, and Vietnam required an effective counterweight to rising Chinese power in Indochina. Recognizing the ongoing conflict between the Soviet Union and China, Vietnam saw its need to secure its interests in a formal, tangible fashion.[9] A treaty with the Soviet Union provided such a vehicle. Vietnam would probably have preferred other forms of demonstrated Soviet support but was willing to accept Soviet pressure for a treaty because of Hanoi's imminent danger. In the treaty, Vietnam extracted some strong security-related language in order to bolster the impression of close Soviet cooperation in the event of a Vietnam-China war.

More important, from the Vietnamese perspective, the Soviet Union provided massive assistance to Hanoi during the Vietnam-China border war in 1978–79, developing a mini-airlift reminiscent of the 1973 support of Egypt.[10] In addition, the Soviet Union did some posturing along the Sino-Soviet border as a reminder to Peking of the potential implications of a truly massive Chinese attempt to overrun Vietnam.

The distinctions between the Soviet-Vietnamese patron-prevalent relationship and the Soviet-Ethiopian patron-centric relationship are instructive. At no time in the former did the Soviet Union attempt to dictate internal Vietnamese policy, nor was the Soviet Union able to deter Vietnamese aggression against Kampuchea. That contrasts sharply with Soviet pressure on Mengistu to establish a vanguard party and to limit the extent of the conflict in the Ogaden. Yet, in both cases, the Soviet Union provided massive arms supplies and engaged in an extensive sea and air lift. In the patron-centric relationship, the Soviet Union was far more successful in controlling its client than it was in the patron-prevalent relationship.

Type III: Influence Parity

Type III relationships vary from types I and II in the goals sought by the patron and in the value the patron places on the client and the relationship. In influence parity relationships, the patron seeks goals of strategic advantage. These are directly related to its competition with other patrons and, because military power remains the fundamental currency of international politics, have the most measurable and tangible reward. The client who has strategic assets is in a most advantageous position to exert influence in the relationship. The patron will go to great lengths, expense, and risk to maintain the relationship.

The client, as in the other relationships described, is faced with a high-threat environment. Again, it will be desperate for patron support and may be willing to grant extensive commitments. The client, too, will value the relationship highly and will work to preserve it, as long as the patron meets the client's perceived essential security needs.

Type III relationships are marked by dramatic and rapid change. Both states will pursue the relationship with vigor and will press for perceived advantages. This type of relationship is the most dangerous in terms of global stability because it provides little room for patron disengagement and provides ample opportunity for rapid escalation. The crisis atmosphere in which the client finds itself contributes to a decrease in rational decision making in the patron, which, in turn, leads to heightened possibilities of patron conflict.

USSR and Afghanistan: September 1979–March 1980

A recent and highly enlightening example of an influence parity relationship is that between the Soviet Union and Afghanistan during the period immediately preceding and following the Soviet invasion.

Increasingly in this period, both the Soviet Union and Afghanistan found themselves desperately needing one another under the highly stressful conditions of a major internal crisis in Afghanistan. The resulting Soviet actions, and their reverberations at the superpower level, again illustrate the dangers of influence parity relationships.

From the Soviet perspective, the elevation of Afghanistan in importance stemmed from three basic sources. First, the insurgency by fundamentalist Muslims and the declaration of Jihad (holy war) by Afghani clerics had established a chain of events disturbingly similar to those that had occurred in neighboring Iran. This was a dangerous precedent in an area contiguous to the Muslim republics of the Soviet Union, where non-Russians were expanding as a significant portion of the Soviet population.[11] Moscow could ill afford to have both Iran and Afghanistan serve as models for its own Muslim minorities and needed to demonstrate Soviet ability and willingness to control Islamic fundamentalism. After considerable private and public gloating over U.S. problems with Iran, the Soviet Union faced a similar prospect with its client, Afghanistan.

Second, there was danger, in the Soviet view, that an Islamic Afghanistan would be vehemently and irrationally anti-Soviet, just as Iran had become vehemently and irrationally anti-American.[12] The dangers of yet another anti-Soviet Islamic state along the Soviet border that could eventually find its way into the U.S. or Chinese orbit presented the Soviet Union with a major strategic danger, one that could not be overlooked.[13]

Third, the Soviet Union saw the Afghanistan situation as an opportunity to extend its control into South Asia. The events in Iran, particularly the seizure of the hostages in November 1979, presented the Soviets with a chance to fish in troubled waters. Soviet divisions in Afghanistan, along with those in the Turkestan and Transcaucasus Military Districts of the Soviet Union, would give Moscow a three-pronged attack option against Iran and a base of operations against Pakistan. In the former case, the Soviet Union recognized that should the Tudeh party or other leftist groups in Iran manage to wrest power from Khomeini, they would probably be unable to control the forces of ethnic separatism. They would probably have to call upon the Soviet Union for assistance. Should Soviet units enter Iran in response to such a plea, it is highly likely that the United States would introduce American military forces into Iran.[14] Soviet divisions in Afghanistan, although a long way from the Khuzestan oil fields, could have a major effect on the outcome of such a Soviet-American conflict in Iran by driving south to Chah Bahar and controlling the Gulf of Oman or by attacking along the Kerman-Sirjan-Hajiabad axis to seize Bandar Abbas.

In addition to dealing with an Iranian contingency, Soviet divisions in Afghanistan could be useful in luring Pakistan away from the United States-China orbit or at least in preventing Pakistan from providing the United States with bases or facilities through which the U.S. could project forces into the region or directly threaten the Soviet Union. Moreover, Soviet airbases in Afghanistan could provide Moscow with a credible strike capability against the expanded United States naval presence in the Indian Ocean.

For these reasons, the Soviet Union goals in Afghanistan became strategic in nature, and Moscow committed combat forces massively for the first time outside of the Warsaw Pact to engage the Afghani Freedom Fighters.

From the Afghani perspective, direct and massive Soviet support became a sine qua non for the survival of the unpopular Marxist revolutionary government. Since its successful coup against Mohammed Daoud in 1978, the ruling Marxist Khalq party had successfully alienated almost every segment of the historically militant and independent Afghani population, particularly in the hinterlands. The declaration of Jihad in the summer of 1979, and the expanding successes of the Freedom Fighters against the disintegrating Afghani army led to insurgent threats to major population centers such as Herat and Kabul.

The political-military situation reached crisis proportions with the overthrow and execution of Nor Turaki by his deputy Hafizullah Amin in September 1979. Amin was a practitioner of the most draconian methods of suppression, which alienated the Soviet leadership, not on moral grounds but for the practical reason that such methods would not work. Indeed, during the four months of Amin's rule, the Freedom Fighters made substantial gains, both in the field and with the Afghani people.

Amin, recognizing that his security situation was worsening, probably invited the Soviet Union to respond with a direct intervention.[15] This behavior, and that of Amin's successor, Babrak Karmal, are the clearest indications of desperate subordination of a client to a patron. The legal government of Afghanistan willingly surrendered itself to a voracious patron in the interest of threat survival.

As in most type III relationships, the price to both the client and the patron was substantial and exceeded expectations. Amin paid for Soviet intervention with the overthrow of his government and his wing of the party and, ultimately, with his life. The Soviet Union paid for its response to Amin's request by becoming bogged down in a costly and initially unwinable war, by losing its few remaining credentials in the nonaligned world, and by undercutting the foundations of détente. The specific Western responses, although criticized as symbolic, were nonetheless significant. The boycott of the Moscow Olympics promised to obviate a significant Soviet propaganda advantage; the economic sanctions, including the grain embargo, produced further strains on the already moribund Soviet economy.

The dangers of a type III relationship are clear in the Soviet-Afghani case. From the client's perspective, it may have gained a chance to control the rebellion, but it lost its independence. From the patron's point of view, it retained and improved its strategic posture in the region, but it stimulated more attention to security within its competitors to its net detriment over the long term.

Type IV: Patron and Client Indifference

In relationships of the first three types, the threat environment for the client is high, and the principal differentiating variable is the patron's goal calculus. In types IV, V, and VI, the client has a low threat environment and, because of this, will be far less willing to accommodate the patron's demands, especially if it is perceived that such accommodation may be internally destabilizing.

A type IV relationship is not greatly valued by either the patron or the client. Neither will make much of an effort to be especially receptive to the demands of the other.

The patron seeks ideological goals in the relationship and direct influence over the client's internal and security behavior. The client, not perceiving an immediate threat, will be generally unwilling to allow the patron this measure of control, unless the patron has some extraordinary contribution to make in nonsecurity areas. Even under conditions of dire economic stress, the client will be normally less receptive to patron demands than it would be under conditions of high threat.[16] The patron, for its part, will not make sweeping offers of vast support or of importance. The relationship, to the patron, is simply not worth the cost. Therefore, such relationships drift along at the edge of relevance and are not usually dangerous to the international community. However, they do often provide the historical bases for rapid transformation into relationships of other types and, therefore, are analytically and substantively important.

USSR-Jamaica: 1978–80

Soviet ties with the Michael Manley regime of Jamaica were type IV in nature; both patron and client displayed marked indifference.

The Soviet Union, during this period, sought primarily ideological goals in the relationship. The shift of Jamaica away from its subservience to Britain and into an espoused socialist mode of development was seen in Moscow as helping to pave the way for other Latin American states to adopt a similar pattern. Consistent Soviet references to Jamaica as a "progressive" state were reminders of the ideological nature of Soviet objectives in the relationship.[17] It was important to the Soviet Union that socialist rhetoric spread beyond the confines of Cuba in order to show that the Soviet model was suitable throughout the region and that Cuba was not simply an anomaly.

At the same time, the Soviet Union had only modest goals of international solidarity and almost no goals of strategic advantage during this period. Although the Soviet Union encouraged Jamaican adherence to Moscow's international line, it was not the centerplace of the relationship.[18] Similarly, during this period, the Soviet Union still conceded the Caribbean to the United States and had little interest in challenging the U.S. directly in one of the few remaining areas in which the United States maintained military superiority. Further, with the availability of Moscow's stalwart ally, Cuba, Jamaica presented only marginal possibilities for enhancing Soviet strategic posture in the region.

The Soviet Union, therefore, had only a moderate interest in the relationship, and these stemmed primarily from its ideological objectives. Moscow had little desire to underwrite the faltering Jamaican economy and be saddled with yet another Cuba. Further, signs indicated that the Manley government would be turned out of office in the general election and that it would be replaced by a regime that would have a less-receptive attitude toward close ties with Moscow. As a result of these factors, Soviet assistance to Jamaica was very modest in nature, consisting of limited economic assistance and advisers on internal security matters.[19]

From the Jamaican perspective, its external threat environment was quite low. As an island nation, Jamaica enjoyed relative immunity from external aggression. Its need for extensive Soviet arms shipments or security assistance was minimal. The principal threat to the Manley regime stemmed not from external invasion but from its internal unpopularity. Manley's philosophical sympathies for the Soviet Union were, in fact, one root of the instability of his government. The United States remained the primary source of economic livelihood for the Jamaicans and had a reasonably good image in the eyes of the Jamaican population.[20] Extensive reliance on the Soviet Union and the domestic and international shift to Moscow's line caused an intensely negative reaction throughout Jamaica.

Manley, aware of this dilemma, could ill afford to entirely alienate his primary trading partner, nor could he risk the ire of his domestic constituency. He had to maintain a respectable distance from Moscow and could accept neither a treaty nor a Soviet military facility, even were any such proposed.

The Soviet-Jamaican relationship endured throughout the period simply because there was no stress applied to it. Neither state had a deeply rooted commitment to the relationship and neither would pay a significant price for its continuance.

Type V: Client Prevalence

Client-prevalent relationships continue to be marked by a relatively low threat environment for the client. The patron's goals, however, have changed into objectives of international solidarity. This elevates the importance of the relationship in the patron's calculus and makes the patron more willing to invest in maintaining the ties. The client, on the other hand, does not have as active an interest in the relationship. The client's demands will be more substantial, its price steeper, and its influence greater than in previous types of relationships.

USSR-India: 1974–80

Throughout much of the decade of the 1970s, the principal Soviet goal in its relationship with India was international solidarity. In the Soviet calculus, India represented a key alliance in the Third World for a number of reasons. First, India was a historical founder of the Non-Aligned Movement and carried with it a substantial leadership role. A close international identification between the Soviet Union and India helped policy Moscow's credentials as an anti–imperialist ally upon which developing states could rely.

Second, India played a pivotal role in South Asia, an area of increasing Soviet interest throughout the decade. A close India-Soviet relationship helped ensure that the United States-Pakistan-Iran alliance did not dominate South Asia. As the United States position in South Asia eroded, Soviet ties with India became an important lever in increasing Soviet political influence in the region.

Third, India provided the Soviet Union with a convenient stalking horse for discrediting the United States in the region. India took the lead on a number of international issues that the Soviet Union recognized the United States could not support. The Indian Ocean Zone of Peace was one such initiative in which India figured prominently. Neither the Soviet Union nor the United States could support the proposal for strategic reasons; yet, when it became apparent that India was taking the lead and that the United States would oppose the effort, the Soviet Union supported the Indian position and laid the blame for the failure of the initiative on the United States.

An exemplary return for the Soviet investment in the relationship was the lack of Indian opposition to the Soviet invasion of Afghanistan.[21] The Soviet Union recognized the important international role India would play in the aftermath of the invasion, and Gromyko hastened to New Delhi shortly after the invasion in order to present the Soviet argument. The success of the Gromyko visit, and the long-term benefits of the relationship for the Soviet Union, became apparent in India's tacit and occasional verbal acceptance of the Soviet action. This had the net effect of helping erode Third World opposition to the invasion and of helping restore quickly the status quo ante in Soviet dealings with portions of the Third World.

Because goals of international solidarity were dominant in the Soviet calculus throughout the period, there was little pressure on India to change its internal makeup. India's quasi-democratic, capitalist society was ideologically anathema to Soviet Marxism—Leninism, notwithstanding brave Soviet attempts to reconcile the two philosophically.[22] Indeed, India was the only state with which the Soviet Union had a treaty of friendship and cooperation that did not espouse, at least superficially, the socialist line.

At the other end of the spectrum of Soviet goals, Moscow had only secondary goals of strategic advantage. Although the Soviet Union made occasional requests to India for basing rights, all were rebuffed, and Moscow had no reason to anticipate any change in Indian attitudes on this issue.

From the Indian perspective, its threat environment was relatively low throughout the period. India emerged from the 1971 war with Pakistan as the clear victor and with the demonstrated support of its superpower patron. The 1974 "peaceful" explosion of a nuclear device and the New Delhi's refusal to allow nuclear safeguards further contributed to its low threat environment. Only China posed a significant threat to India, and the Chinese were relatively quiet throughout the period.

Because of this low threat environment, India was able to deal with the Soviet Union from a position of political strength. India looked to Moscow as its principal arms supplier but demonstrated its willingness to shop elsewhere, should the Soviet Union prove too demanding in its political or economic terms.[23]

Again, the nature of the relationship was clearly demonstrated following the invasion of Afghanistan. In exchange for Indian international solidarity, the Soviet Union extended New Delhi a $2.1

billion arms deal on highly concessionary terms.[24] In the absence of Indian political support, it is doubtful that such an agreeement would have been offered.

Thus, the Soviet-Indian ties during this period were type V in nature, with India playing a dominant role. As long as India continued to support the Soviet Union in the international arena, the Soviet Union was willing to provide India with extremely high levels of security assistance and was willing to accept India's ideological orientation.

In the case of the USSR and India, the patron provided enormous amounts of military hardware at exceptionally favorable rates to the client in return for support by the client on international issues. The type V nature of the relationships provided greater opportunity for both client states to exert influence over their patrons.

Type VI: Client-centric

The most dangerous form of patron-client state relationship, from the perspective of the international community is the client-centric relationship. It is a relationship of this form in which the client exerts the maximum amount of influence over the patron and has the greatest access to the patron's military and political resources. These are the relationships that run the greatest risk of escalating into a larger confrontation, perhaps involving the superpowers themselves in direct conflict.

Such a relationship occurs when the patron has a major strategic advantage as its goal. Through its ties with its client, the patron can gain a significant and measurable advantage over its patron opponents. In a world of increasing competition between patrons, and especially between superpowers, a client that yields a significant strategic advantage over other patrons is highly prized. The client, for its part, does not face a major threat environment and is, therefore, more able to negotiate with the patron from a position of strength. This is even more pronounced if the client can present the impression of being available to the highest patron bidder.

USSR-Vietnam: 1979–80

The relationship between the Soviet Union and Vietnam underwent changes along the two basic continua that altered the relationship from the patron-prevalent type mentioned above to a client-centric relationship. The first change was the reduction in Vietnam's threat environment, following the withdrawal of the Chinese and the beginnings of talks between the two states. Although China remained a significant threat, and problems continued with Kampuchea, the threat environment from Hanoi's perspective became relatively low. No longer completely distracted by the Chinese incursion, Vietnam was able to continue its quest for regional domination, confident of Soviet support.

The Soviet Union, for its part, began to see Vietnam in terms of strategic advantage, primarily because of the rising importance of the Soviet naval presence in the Indian Ocean. During the Indian Ocean Arms Limitation talks, the Soviet Union had hoped that an agreement would relieve Moscow from competition in the region, at least until it was in a better position to compete. But, as events developed, the United States withdrew from the talks and stepped up its naval presence to unprecedented levels.[25] In order to pose a credible threat to the sea lanes through which oil flowed to the West and in order to counter U.S. naval presence, the Soviet Union was obliged to commit larger forces into the region. But support from Vladivostok or the Black Sea was difficult at best. Therefore, ports in Vietnam became increasingly valuable to Soviet Indian Ocean operations.

This, in turn, escalated the importance of Vietnam to the Soviet Union beyond ideological or international solidarity issues. Now, strategic advantage became involved, and the Soviet Union became more accommodating to Vietnamese adventurism in the region.

Tacit Soviet support of Vietnamese incursions into Thailand was indicative of the change in Soviet attitude. Although the United States and the Association of Southeast Asian Nations (ASEAN) were willing to tolerate Vietnamese actions in Kampuchea, a strong reaction could be anticipated to threats against

Thailand. Vietnamese aggression against Thailand could well bring about a reintroduction of U.S. forces to the region, this time in an alliance with China. This the Soviet Union could ill afford. Yet, there was little indication of any actual restraint imposed by Moscow on Vietnam to control such incursions. The promise of Soviet bases at DaNang and Cam Ranh Bay was so enticing to Moscow that it was willing to tolerate highly dangerous Vietnamese activity along the Thai border.[26]

Again, it is not difficult to project circumstances in this relationship in which direct superpower conflict could occur. Vietnam's confidence in Soviet support, regardless of how Vietnamese ambitions related to Soviet desires, had the potential for Vietnamese control over Soviet military assets to an unprecedented degree. The flow of influence in the relationship was such that the Vietnamese were in a position to involve the Soviet Union on a ladder of escalation that far exceeded Soviet gains from the relationship itself.

CONCLUSION

In all forms of patron-client state relationships, the key prizes are control over resources and the flow of influence. The basic dimension that yields qualitative assessments of the danger of patron-client relationships to the international system is the extent to which the client can control the military and political resources of the patron. The greater the client's control, the greater is the danger of conflict on a global scale. The patron will give up the most control when it sees strategic value in the relationship; the client will control most when it has a modest threat environment and is, therefore, not desperate for patron support.

Patron-client state relationships, then, cannot be adequately understood by treating them as a uniform whole. Rather, an examination must be made of the goals and threats that underlie them, always considering that these bilateral ties are really instruments in an increasingly competitive patron environment.

NOTES

1. Richard B. Remneck, "Soviet Policy in the Horn of Africa: The Decision to Intervene," paper prepared for the Conference on the Soviet Union and the Third World, U.S. Army War College, Strategic Studies Institute, September 1979, Carlisle Barracks, Penn., p. 27.

2. *Washington Post*, April 25, 1978, p. A-17.

3. U.S. Foreign Military Sales (FMS) agreements went from $138 million in 1976 to nothing by the last months of 1977. *Foreign Military Sales and Military Assistance Facts*, (Washington: Defense Security Assistance Agency, 1978), p. 2.

4. A. G. Noorani, "Soviet Ambitions in South Asia," *International Security*, Winter 1979–80, p. 32.

5. Frederick H. Hartman, *The Relations of Nations*, (New York: Macmillan, 1978), pp. 333–36.

6. The importance the United States attached to the U.N. vote on Afghanistan is ample evidence of this. The vote was the source of great pride to the Carter administration.

7. Daniel Zagoria, "Soviet Policy and Prospects in East Asia," *International Security* Fall 1979, p. 72.

8. William S. Turley and Jeffrey Race, "The Third Indochina War," *Foreign Policy*, Spring 1980, p. 93.

9. Ibid., p. 95.

10. Douglas Pike, "Communist vs. Communist in Southeast Asia," *International Security*, Summer 1979, p. 27.

11. Ann Sheehy, *The National Composition of the Population of the USSR According to the Census of 1979* (Washington: Radio Liberty/Radio Free Europe Paper, 1980).

12. Iran was a source of moral and material support for the Afghani Mujahdeen. *Washington Post*, March 22, 1980, p. 1.

13. This was the basic Soviet justification for the invasion. *Foreign Broadcast Information Service, USSR International Affairs*, March 13, 1980, p. D-1.

14. Indeed, Carter promised exactly that in his State of the Union Address in 1980.

15. The first Soviet airborne division may have arrived in Kabul a full week before Amin's execution. John C. Griffiths, *Afghanistan: Key to a Continent* (Boulder, Col.: Westview Press, 1981), p. 149.

16. Economic problems do not have the sense of urgency that surrounds a direct military threat; Third World states tend to treat such problems more casually and, in some ways, with a more rational approach than they do military problems.

17. The defeat of the Manley government in October 1980 ended the Soviet-Jamaican relationship, demonstrating the tenuousness of a type IV relationship.

18. Jamaica voted for the U.N. resolution condemning the Soviet invasion of Afghanistan, for example. *New York Times*, January 15, 1980, p. 8.

19. By the end of 1978, for example, Soviet aid to Jamaica consisted of only $70 million for a cement plant. *Financial Times of London*, December 8, 1978.

20. This remained true even at the height of Manley's relationship with the Soviet Union. *New York Times*, January 21, 1978, p. 19.

21. India abstained in the U.N. vote. *New York Times*, January 15, 1980, p. 1.

22. Aryeh Yodfat, *Arab Politics in the Soviet Mirror* (Jerusalem: Israel Universities Press, 1973), pp. 2, 6–8, 14–15, 24, 62–64.

23. India, for example, purchased the British JAGUAR aircraft. Pushpindar Singh, "JAGUAR Enters Service With India's Air Force," *Armada International*, February 1980, pp. 71–82.

24. *New York Times*, May 29, 1980, p. 16.

25. By 1980 the U.S. naval presence in the Indian Ocean had grown to two aircraft carrier battle groups on station all the time.

26. Robert L. Pfaltzgraff, Jr., "China, Soviet Security and American Policy," *International Security*, Fall 1980, p. 29.

CHAPTER 4

THE DYNAMICS OF THE RELATIONSHIP

Up to this point, our discussion of patron-client state relationship has focused on the goals that shape the nature of the relationship and determine the flow of influence. The examples treated show particular types of relationships, and goal structures, at specific times.

History demonstrates that patron-client state relationships are not static in nature. The purpose of this chapter is to examine the dynamics of patron-client state relationships and the mechanisms by which they change.

CHANGE IN THE RELATIONSHIP

Understanding change is important for both analytical and practical reasons. From an analytical perspective, the richness of patron-client state relationships cannot be appreciated without grasping the fundamental role change plays. It is change that introduces the basically destabilizing characteristics that will be discussed in more detail below. Change is as basic a quality of patron-client state relationships as the goal structures that give them purpose and form. It is also change that, in the context of crises, presents the greatest dangers for the international system.

From a practical perspective, understanding the mechanisms of change presents some real policy options for decision makers. The shaping of the relationship by either the patron or the client is facilitated by understanding how and why change occurs. Similarly, the erosion of competitor patron-client state ties may be encouraged and assisted by understanding and exploiting change mechanisms. Simple examination and comprehension of the goals each state seeks in the relationship is insufficient for this purpose.

An appreciation of change in patron-client state relationships rests upon an examination of two subordinate questions: Why does change occur, and how does change occur.

Reasons for Change

Patron-client state relationships are rooted in the goal structures of each state and in the nature of the client's threat environment. Changes in any of these areas or changes in the international milieu in which such relationships are set can bring changes in the relationship itself.

Internal Goal Structure

Since the nature of the relationship is determined by the goals each state seeks, changes in these goals will alter the basic objectives sought in the relationship and the value the relationship has for each state. This, in turn, will change the nature of the relationship itself. Internal goal structures can change in a variety of ways, several which are outlined below.

First, the patron's military capabilities may change vis-à-vis its opponents. The client may thus assume a more or less substantial role in the patron's strategic advantage calculus. This type of change is derived from advances in the patron's technology, increases in the patron's force structure, or changes in the strategic posture of the patron's opponents. It is usually independent of developments within the client. Several examples can serve to illustrate this point.

In its seemingly random expansion of its force structure during the 1970s, the Soviet Union committed an increasingly significant naval presence to the Indian Ocean, an area that previously had been largely neglected in Soviet plans primarily because of the limited size of its navy. In the 1970s, however, the Soviet Navy grew substantially in size and scope.[1] And, during the same period, Soviet naval presence grew in the Indian Ocean.[2] The Indian Ocean region presented a particularly tempting target for Soviet expansionism; not only was the West increasingly dependent on the oil that flowed from and through the region, there was also no substantial Western military presence with which the Soviet Union would have to deal or to which the still fledgling Soviet Navy could be unfavorably compared.

But the expanding Soviet Indian Ocean Squadron (SOVINDRON, in U.S. Navy parlence) presented a new set of onshore problems that led to changes in Soviet relations with littoral states. As long as the Soviet Union maintained a relatively modest presence in the region, it could replenish and support its vessels largely through a train of support vessels. This, in turn, reduced the requirement for regional host-nation cooperation. However, as SOVINDRON expanded, the sea train logistical concept was no longer entirely adequate. Onshore facilities for repairs, for logistics, and for crew liberty became a sine qua non for the larger Soviet naval presence.[3]

In addition, as the Soviet Union committed more valuable naval assets, such as guided missile cruisers, to SOVINDRON, protection from U.S. attack submarines became more important. Like the U.S. Navy, the Soviets rely primarily upon anti-submarine warfare (ASW) aircraft to detect, track, and attack enemy submarines. But the Soviet IL-38 MAY aircraft, a mainstay of Soviet ASW capabilities, needed to be staged from regional facilities; their range was not sufficient to allow them to be staged from the USSR, nor could the Soviet Union rely upon blanket overflight permission from Iran, Pakistan, or Turkey.

Therefore, as the Soviet Navy began to challenge the West in the Indian Ocean, the value of relations with littoral states expanded substantially in Soviet calculations. The crisis in the Horn of Africa in 1977–78, during which the Soviet Union doubled its normal naval presence deployed in the Indian Ocean, brought the support problem sharply into focus. After that point, Soviet relations with Ethiopia, the People's Democratic Republic of Yemen (PDRY), and to a lesser extent, Madagascar, changed dramatically. No longer were these states simply ideological burrs in the saddle of the West. Now they acquired genuine strategic value for the Soviet Union, and the relationships between Moscow and Addis Ababa, and Moscow and Aden, moved from the left to the right (see Figure 2.3), with a concomitant increase in Soviet commitment to each state.[4] The relationships between these littoral states and the Soviet Union changed because Soviet capabilities began to allow it to compete with the West and particularly with the United States in the Indian Ocean region, with potentially significant returns.

The Soviet relationship with Cuba during the 1960s shows this point as well, with the relationship changing form and character several times. "Missile gap" rhetoric notwithstanding, the Soviet Union was at a distinct disadvantage during the early years of the decade of the 1960s. The Soviet inability to counter the missile and bomber edge enjoyed by the United States with ICBM systems of its own led to a greatly expanded strategic advantage role for Cuba. Medium Range Ballistic Missiles (MRBM) and Intermediate Range Ballistic Missiles (IRBM), both of which the Soviet Union possessed and could manufacture with relative ease, could be emplaced in Cuba and could provide a quick solution to Soviet strategic inferiority.[5] Therefore, Cuba became far more valuable to the Soviet Union that it had been in the past.

However, the forceful U.S. response to the emplacement of the missiles and the subsequent Soviet development and deployment of an ICBM force based in the Soviet Union, substantially reduced the value of Cuba in Soviet calculations. As a result, Soviet-Cuban relations became more strained in the mid- and late 1960s.[6]

The resurgence of close Soviet-Cuban relations in the mid-1970s was tied to the new strategic role that Cuba assumed. The high utility of Cuban troops as surrogates for Soviet forces in the increasingly suspicious Third World elevated Cuba once again in Soviet calculations. The use of Cuban troops allowed the Soviet Union to exploit local instabilities and to undermine Western interests without risking direct confrontation by committing its own forces. The curious credibility that Cuba enjoyed, particularly in

Africa, made it an invaluable adjunct to Soviet expansionism in that part of the world.[7]

Obviously, there are many other factors that helped shape the complex relationship between the Soviet Union and Cuba during the decades of the 1960s and 1970s. However, the underlying cause of change in the relationship was the strategic value that the Soviet Union placed on Cuba. As this calculation changed, Soviet willingness to pay the economic and political price for support of Cuba and to maintain the relationship changed as well.

Goal structures change not only within the patron but also within the client. The principal change that occurs in the client's goal structure is in its perception of its threat environment. As threats to the client change, the client's willingness to contribute to the relationship will change as well. As a manifest internal or external threat emerges that endangers the client, it will become far more willing to make major concessions to its patron. As the threat abates, the client will attempt to lessen its commitment and will try to mitigate the concessions it made during the high-threat period. The client will attempt to drive the relationship toward types IV, V or VI (see Figure 2.3) when not pressed directly by a major external threat. These types of relationship imply the level of commitment the client desires. Again, several examples illustrate this point.

The regime of President Ali Salih ʿAbdullah in the Yemen Arab Republic (YAR) was noted, during the period 1978–80, for its instability. At almost any moment, it seemed, the variety of challenges to Salih stemming from the National Democratic Front, tribal rivalries, Saudi and PDRY neighbors, and the superpowers themselves would bring Salih down. Salih remained in power through a delicate balancing act in which he remained uncommitted to all sides, placating and resisting at alternate turns. Every side remained convinced that the successor to Salih would be more difficult to control. So Salih survived by avoiding, or at least balancing, commitments.

However, in March of 1979, the NDF with PDRY backing chose to make a bid to overrun the southern portions of the YAR. Faced with a major crisis originating in the Soviet-backed NDF-PDRY coalition, Salih turned to the United States. His reluctance to make commitments and to appear aligned with any camp disappeared quickly in the heat of battle. He invited United States resupply, advisers, and military presence in the YAR, even though by doing so he increased, in the long run, the likelihood that his regime would fall.[8]

The United States responded quickly and, when measured in post-Vietnam terms, dramatically by promising some $390 million worth of military equipment and by using the emergency provisions of the Arms Export Control Act to circumvent normal Congressional notification procedures.[9] The United States had a rare opportunity to use the conflict and Salih's increased willingness to accomodate U.S. desires to establish an American presence in the YAR, to force the expulsion of Soviet advisers, and to balance Soviet presence in Ethiopia and PDRY. Thus, changes in Salih's perceived threat environment, coupled with expanding Soviet activities in the region, drove the relationship from type IV to type III.

In the aftermath of the conflict, which turned out to be quite small and inconsequential, Salih was quick to try to go back on his promises. He accepted the Iraqi plan for negotiations to unify the two Yemens (Saudi opposition notwithstanding), and he began a rapproachment with the Soviet Union. Salih did not renounce his American connection, and he continued to press for U.S. assistance. But his willingness to make the kinds of concessions most desired by the United States greatly diminished as the threat to his regime abated. Thus, he drove the relationship from type III back toward row type VI in an effort to reduce his commitment to the United States and to Saudi Arabia.[10]

Reducing Angola's concessions in the Angola-Soviet relationship proved more difficult for the regime of Agustinho Neto. During the high-threat period of his consolidation, while the issue of his control remained very much in doubt, he willingly acceded to Soviet and Cuban demands for extensive control over Angola's internal affairs. However, as the threat from the Uniao Nacional de Independencia Totale de Angola (UNITA) temporarily stabilized in the south, Neto tried to begin to move away from strict adherance to the Soviet line. Recognizing that Western, and expecially Portuguese, support was essential to his ailing economy, Neto made a number of gestures to the West in December, 1978. Soviet and Cuban presence were well ensconced by this time, however, and Neto found it extremely difficult to effect closer relations with the West and to deny the Soviet Union staging rights for IL-38 MAY ASW aircraft and for naval vessels. Thus, although the threat environment dictated a movement of the relationship from type III to type VI, the Angolan commitment to the relationship remained at a type III level, propped up artificially by Soviet and Cuban control over many of the internal levers of power in Luanda.[11]

Internal goal structures also change as a result of change in leadership within both patron and client. New patron leaders may direct a reassessment of the worth of the client, particularly vis-à-vis other priorities. The election of Jimmy Carter to the presidency in 1976 spelled the end of the United States-Nicaragua relationship because of the new president's commitment to a new set of priorities, led by human rights.

Even more significantly, changes in leadership in the client state will affect its relationship with its patron. This is particularly true if the change occurs through an ideological revolution. In 1969, Muhammar Qaddafhi seized power in Libya, and the nature of the U.S.-Libyan relationship changed drastically. Qaddafhi promptly assumed a radical, anti-Western international posture, ejected the British from the country in which they had maintained a continuous presence since 1882, and terminated the U.S. rights to Wheelus Air Force Base. The history of excellent Libyan relations with the West was to no avail when faced with revolutionary fervor. Similarly, the Soviet position in Indonesia disappeared almost overnight with the downfall of Sukarno. His successor, General Suharto, would tolerate no dealings with the Soviet Union and quickly ended the Indonesian-Soviet relationship. Even the largest and most nominally democratic of the Third World states, India, is not immune from such dramatic changes. When Desai was elected in 1976, the close Indian-Soviet relationship suffered an almost immediate chill. The subsequent reelection of Indira Ghandi restored close Soviet-Indian ties.

Patron-client state relationships are highly susceptible to dramatic change when the internal leadership in the client changes because of the highly personal nature of leadership in client countries. Often, the policies of a government in the Third World are synonymous with the views and style of an individual leader. Ties with patron states are usually the product of one leader's desires and are rarely institutionalized. This is true regardless of the ideological leanings or the constitutional format of a client state. Thus, patron-client state relationships can often not stand the political strains imposed on them when Third World client regimes change. Patron states, on the other hand, are normally marked by more stable political systems and a more rigid set of political values. Associations between patrons and other states, including clients, are more often products of national interests, as opposed to personal whims of individual leaders. As such, these relationships can better withstand the turmoil of changes in internal leadership.

This point was also illustrated in the Soviet relationship with the PDRY. In Ismail, the Soviet Union found a leader who was ideologically compatible and admired the Soviet system. Throughout his tenure in power, Ismail forged close links with the Soviet Union and allowed Moscow to become the dominant force in South Yemeni politics. However, Ismail was plagued with a number of internal problems, not the least of which was his ethnic origin. He was born in North Yemen, a fact greatly resented by the southerners who dominated the rest of the party and government.

The Soviet Union recognized that its power base in PDRY in 1979 was dependent upon the survival in power of party leader Ismail. It is highly likely that the Soviet Union was responsible for forging the compromise with Ismail's opponents that, in August 1979, kept Ismail in control. However, the compromise collapsed some nine months later, and a duumvirate of Al-Hasani (Ali Nasir) and Ali Antar forced Ismail out of the party and government.[12] Because both new leaders were committed Marxists and had no alternatives to Soviet support, the relationship endured. However, both Al-Hasani and Ali Antar worked to erode Soviet influence in the domestic affairs of the PDRY.[13]

The highly personal nature of many patron-client state relationships is yet another factor that adds to their inherent instability and that contributes to change over time. In addition, this factor provides opportunities for international mischief on the part of the patron's competitors who may seek to undermine a relationship by fostering changes in the internal leadership of the client.

The International Milieu

Change is not only caused by changes in the internal goal structures of the patron and the client. It is also brought about by changes in the international milieu. The distinction between relationship changes caused by internal goal structure alterations and those caused by changes in the international milieu is subtle but analytically important. The former occurs basically independent of the environment external to the relationship; the latter, by definition, is a direct function of that environment.

This distinction is important for two reasons: first, it shows the conceptual and practical duality of patron-client state relationships. On the one hand, such relationships are bilateral in their change dynamics and can be manipulated on a strictly bilateral basis. On the other hand, patron-client relationships are inextricably enmeshed in the international political system, subject to the forces that act thereupon. Second, it has significant policy implications for those decision makers who forge, maintain, or dissolve patron-client ties.[14]

As a general rule, patron states are more affected by changes in the environment than are their client partners. Since patron-client state relationships are, at their most basic level, extensions of intra-patron competition, the value the patron places on the relationship will be highly susceptible to its relations with other patrons. For the patron, a particular relationship it may have with a client is but one element in a network of relationships and interactions that make up its competitive environment. Changes anywhere in that competitive environment produce ripples throughout the entire system that, in turn, often lead to major changes in specific relationships.

Client states, too, react to changes in the international environment. This is particularly true in cases where the client has ties with more than one patron. However, because the network of client relations with other states is so much smaller than that of its patron, it is more immune to changes in the international milieu.

Changes in relationships generated by the international environment are common. The Soviet invasion of Afghanistan in December 1979 caused the United States to begin a series of major changes in its relationships with other states in the region. These states quite suddenly assumed a much more important role in the U.S. strategic advantage calculus than they had enjoyed in the past. More importantly, the Soviet invasion turned the philosophical tide of the Carter administration away from détente and toward a more confrontational policy with the Soviet Union. This, in turn, translated into an increased willingness by the United States to engage the Soviet threat directly with U.S. military power, particularly in the Persian Gulf region.[15]

For the states in the region, this meant that the nature of their relationships with the United States changed dramatically. Weapons requests that had been denied to the Gulf states of Bahrain, UAE, and Qatar before the Soviet invasion were approved afterwards. This involved the de facto repeal of the Lower Gulf Arms Transfer Policy, a much-valued program of the U.S. arms control community. In addition, the United States entered into quasi security pacts with Oman, Kenya, and Somalia, initiatives that were not contemplated before the Soviet Union moved into Afghanistan. This reversed the decade-long policy of retrenchment and isolation that marked United States ties with these regional states. All these actions were brought about not by the internal factors mentioned in the earlier section of this chapter but rather by changes in the international environment in which these relations were set.

In a similar fashion, Soviet relations with Syria changed in the 1979–80 time period because of factors outside of the relationship itself. The Camp David Accords with the subsequent 1979 peace treaty between Egypt and Israel effectively froze the Soviet Union out of a major role, save for that of nay-sayer, in the most volatile issue in the Middle East. The invasion of Afghanistan arrayed almost the entire Islamic world against the Soviet Union, exposing the insincerity of Moscow's claims to be the champion of the Third World. At the same time, the United States assumed a position of regional favor, notwithstanding the Palestinian issue, that threatened to allow it to shut Moscow out of the region altogether. In this environment, Syria became of much greater importance to the Soviet Union; indeed, Syria became Moscow's sole reliable ally in the region. The Treaty of Friendship and Cooperation, signed in September 1980, codified the drift of the relationship. Syria proved to be useful to Moscow during the subsequent Iraq-Iran war as a conduit for influence in Tehran.

Changes in the international system and especially in the patron's relationships with other patrons, can produce profound changes in individual patron-client state ties, even though there appears to be little relationship between the changes themselves. Again, this factor reflects the basically complex and dynamic nature of patron-client state relationships.

Change in patron-client state ties occurs, then, through two basic mechanisms: changes in the internal goal structures of patron and/or client, and changes in the international milieu in which these relationships are enmeshed. These are, quite obviously, two volatile variable areas that change with great frequency. We would expect, therefore, that patron-client state relationships would change all the time as well. And this is clearly the case. Moreover, such a tendency for change is reinforced by the deliberate manipulation of change mechanisms by both patron and client.

How Change Occurs

Patron-client state relationships are highly dynamic and constantly being redefined for the reasons outlined above. Although the types and directions of change are many and varied, it is useful to visualize two discreet change mechanisms that derive from the conditions leading to the change. Change in patron-client state relationships occurs either through noncrisis or crisis negotiations. *Negotiations* is a particuarly apt term because it implies the sense of give and take that characterizes relationships of this nature. The distinction between noncrisis and crisis negotiations is of critical importance because of the impact each has on the relationship and, potentially, on the international system.

Noncrisis Negotiations

Seldom, if ever, is a patron-client state relationship stable over long periods of time. Given that, at the most basic level, the states have different and conflicting objectives in the relationship, there will be ongoing efforts by each state to extract the maximum advantage possible from the other. During negotiations in noncrisis environments, each state will use a variety of techniques to obtain desired concessions from the other at a minimal expense. In addition, each state will tend to integrate the relationship more rationally into its overall global or regional desires, often to the detriment of its partner.

Noncrisis negotiations are typified by a series of overt or implied proposals and counterproposals between the two partners in the relationship. Change in the basic relationship during such times usually occurs in an incremental fashion with the shape of the relationship ill defined and murky, even to the partners themselves. Neither side is always sure of the extent of commitment of the other.

The patron, operating in an environment of intra-patron competition and having determined what contributions the client can make (ideological, international solidarity, or strategic advantage), can further determine the extent to which it is willing to go to establish or maintain the relationship. As in any market situation, the patron will try to achieve its objective at a minimal cost; if it can pay an ideological price for a strategic advantage goal, it will do so. The patron is reasonably confident that it has the time and the patience to wait the client out and to obtain the most favorable agreement possible.

The client sees its position in similar terms. Noncrisis negotiations, which should not be confused with a low-threat environment, allow the client to negotiate in a more deliberate manner. These conditions also permit the client to examine other patron alternatives, either for genuine reasons or to demonstrate to its patron the tenuousness of the relationship. In addition, the noncrisis situations allow the leadership in the client to condition its constituency to changes in the relationship or to plumb the degree of opposition that various relationship concessions might elicit.

Noncrisis negotiations also allow each side to feel more confident that it knows what the other is up to. Although the final price of the relationship is unclear to either side, the process by which that price is being determined is thought to be understood. This is typified by reasonably firm objectives held by each side, by a series of negotiating steps designed to achieve these objectives, and by limits established and adhered to on the extent of commitment each side is willing to give.

Clearly, the objectives each side seeks in the relationship may change during noncrisis negotiations. When this occurs, new tactics and specific objectives are established, and the negotiations continue.

Several examples illustrate noncrisis negotiations and show the deliberate, incremental nature that is characteristic of this sort of patron-client state relationship change mechanism.

The relationship between the United States and Somalia between 1977–80, took place in a basically noncrisis environment. As such, it was marked by a clear understanding by each state of its objectives and by the offers, counteroffers, and diplomatic chicanery that typify such negotiations.

From 1977 through 1979, the relationship was basically type I in nature. Somalia experienced a reasonably high-threat environment stemming from its irridentist activities in the Ogaden region of Ethiopia and, to a lesser extent, in northeast Kenya.[16] After the 1977 abrogation of the Treaty of Friendship and Cooperation with the Soviet Union, Somalia's President Siad Barre was left with no source of security. Faced with a strong Ethiopian-Cuban threat in the Ogaden, Siad tried to get the United States to replace the Soviet Union as his principal provider of weapons and, more importantly, to guarantee Somalia's territorial integrity. Thus, Siad found himself in a high-threat environment, although not a crisis.

The United States, for its part, had no real strategic interests in Somalia at this time. The Carter administration was still committed to dialogue with the Soviet Union and to demilitarizing the Indian Ocean region.[17] Moreover, the Somali regime was repugnant to the human rights advocates within the U.S. government. Further, Siad had largely brought his troubles on himself by launching the war with Ethiopia in the first place. U.S. objectives in the relationship, then, were primarily ideological in nature. The United States sought an end to Somali aggression in the Ogaden, Somali support for the Western Somali Liberation Front (WSLF) and Somali claims on Djibouti and northeast Kenya. More significantly, the United States sought changes in the Somali system of government, particularly as applied to ethnic and tribal minorities.

Thus, both the United States and Somalia were reasonably certain of their respective objectives. The noncrisis negotiations that followed served to shape the relationship and to establish the limits of commitment of both sides. As might be expected in a type I relationship, the patron remained rather firmly in control of its commitment while the client offered a variety of inducements, largely to no avail.

Initially, after the Soviet expulsion in August 1977, Siad thought that the United States would respond immediately. He had witnessed the dramatic turn of fortunes for his friend Anwar Sadat vis-à-vis the United States and felt that his actions would lead to similar American sympathies for Somalia. Siad's strategic concept of the region, one supported by Sadat, saw the Horn of Africa as particularly significant. Ejection of the Soviet Union from the port of Berbera was, in Siad's view, a monumental gesture to the United States because it freed the Western oil lanes from the threat of Soviet interdiction, and it opened the Red Sea to the Indian Ocean once again. From a regional perspective, Siad's logic was uncontestable. He could, therefore, not understand the failure of the United States to seize the opportunity his expulsion to the Soviet Navy offered.

However, the United States was pursuing a different course, in which Somalia and Berbera played but a minor role, in its bilateral dealings with the Soviet Union. If the negotiations with the Soviet

Union on the Indian Ocean were successful, and in late 1977 there was no reason to suppose that they would not be, bases along the littoral would be severely constrained, and any U.S. investment in Berbera would probably have to be undone. Besides, a U.S. effort to replace the Soviet Union in Berbera would be contrary to the spirit of the Indian Ocean talks.[18]

At the same time, Siad became harder pressed by the Ethiopian-Cuban counteroffensive in the Ogaden, and it became apparent that Somalia itself might be attacked. Siad then changed his negotiating tactics and pledged to reduce his support to the still-active WSLF in the Ogaden. This elicited a modest response in Washington; the United States sent a warning to the Soviet Union not to violate Somali territorial integrity.[19] Yet no further support was forthcoming, although there was debate within the U.S. government on sending an aircraft carrier battlegroup to the region to demonstrate U.S. concern.[20] The U.S. policy on arms transfers to Somalia did not change.

As the situation in the Ogaden stabilized in 1978, due primarily to the diversion of Ethiopian and Cuban attention to the rebellion in Eritrea, Siad once again changed his approach. He began a cautious rapprochement with the Soviet Union, one that was designed far more for U.S. consumption that it was a serious effort to resume the Somali-Soviet relationship. Again, Washington remained unconvinced, although, in early 1979, the United States did dispatch a team to Mogadishu to negotiate the necessary memoranda of understanding that would allow Somalia to make limited purchases of U.S. military hardware.[21]

Finally, in mid-1979, Siad played an effective card. He offered Berbera and Mogadishu again, this time in more explicit terms, to the United States as bases. He also pledged again to withdraw regular Somali troops from the Ogaden and to reduce his support to the WSLF.

With agonizing slowness, the United States responded. The international environment had changed over the course of two years since 1977, and the United States began to see the need for an expanded naval presence in the Indian Ocean. Under these conditions, Somalia became important not for ideological reasons but for reasons of strategic advantage. Siad's suggestions on Berbera and Mogadishu coincided with the willingness of the United States to consider moving to a type III relationship, as opposed to the type I that Washington had supported up to that point. Thus, while Siad's objectives in the relationship remained unchanged, U.S. objectives changed markedly. A new set of negotiations, still in a noncrisis environment, began with the dispatch of a high-level U.S. team to the region to discuss basing options with Oman, Kenya, and Somalia.[22]

The pace and tone of the negotiations, at least from the U.S. standpoint, were established by the rapid conclusion of agreements with Oman and Kenya.[23] These agreements put Siad in a less advantageous position, for although the advantages to the United States of access to Berbera could not be entirely replicated by access to Oman's Masirah or Mombasa in Kenya, there was some degree of strategic overlap. Therefore, Berbera, while still strategically valuable, was expendable. The United States was willing to increase its support for Somalia in exchange for an access agreement but only at a rather modest economic and political cost. The United States did not want to "lead with dollars" and did not want to become enmeshed in Siad's quagmire in the Ogaden.

By mid-1980 both states recognized that the relationship would be fundamentally type III in nature; that is, the United States abandoned its efforts to control Siad's actions in the Ogaden and concentrated on access to Somali facilities, while Siad saw Somali security increasingly tied to a strong relationship with the United States. The degradation of the Somali equipment, which had been manufactured in the Soviet Union, the continued presence of large Cuban units in Ethiopia, and the slowdown of rebel activity in Eritrea all heightened Siad's perception of a high-threat environment. With both states recognizing that a type III relationship was at hand, the issue became the negotiating of the price each would pay. This is worth some mention, if only to show the techniques patron and client may use in order to effect an agreement.

Early in the negotiations, Siad let it be known that he sought some $2 billion in economic assistance, and substantial U.S. security guarantees, as the price tag for an access agreement. In order to avoid a personal rejection of this figure, which he probably recognized was absurd, Siad communicated his proposal through personal intermediaries and through selective leaks to the press.[24] Siad sought to focus U.S. attention on his exorbitant demands for financial assistance and away from his ongoing activities in the Ogaden.

The United States, however, having already initiated agreements with Oman and Kenya, and having begun discussions with Egypt for the use of Ras Banas/Berenice, was in no mood to shift the focus of the discussion. Siad, through intermediaries, was informed that the $2 billion figure was not worthy of argument and that, when Somalia was ready to talk in realistic terms, the United States would reopen the negotiations. The United States allowed that it was prepared to discuss modest amounts of economic and security assistance but that, under no conditions, would this assistance be the primary ingredient in a security relationship. To reinforce the point, Secretary of State Muskie announced that the United States was rethinking its position on seeking access to Somalia.[25]

Siad then dropped his monetary demands from $2 billion to the $40 million in security assistance over two years informally promised by the U.S. negotiators. He also again announced that all Somali regular troops would be withdrawn from the Ogaden. He further announced that the real value to Somalia in a security relationship with the United States would be a security guarantee, rather than financial assistance.[26] This position was ultimately acceptable to the United States, and a formal access agreement was signed in September 1980.[27]

However, this did not end the process of negotiations as far as the relationship itself was concerned. The perceived threat to Somalia from Ethiopia grew in Siad's eyes, driving the value of the relationship up. At the same time, budgetary constraints and a growing U.S. interest in Ras Banas were reducing the value of the relationship in the U.S. calculus. Therefore, Siad was in a hurry to begin implementing the terms of the agreement, and the United States stalled, sought redefinition, and established new conditions under which the agreement would be honored.[28]

Throughout this period, each side sought to extract maximum concessions from the other at a minimum cost. The United States believed that it had extracted a pledge from Siad to remove Somali troops from the Ogaden and, effectively, to renounce his irridentist claims in the region. In addition, the United States believed it had carte blanche to do almost anything with Berbera and, to a lesser extent, with Mogadishu. Siad, for his part, believed that the United States would eventually become Somalia's security guarantor and that Somalia would enjoy extensive U.S. security assistance and a free hand in the Ogaden. These obviously contradictory positions resulted in the extensive negotiations after the formal access agreement was signed.

The relationship between the United States and Somalia during this period typifies the sorts of negotiations that go on between patron and client in noncrisis periods. The relationship was certainly not static during this period; indeed, it progressed from type I to a modest type III without the catalytic element of a crisis. However, even in the face of an expanding threat environment, the negotiations were relatively relaxed, with each side pursuing its changing objectives in a realistic fashion. Neither side panicked and offered concessions that it later regretted. Neither side was willing to expand its commitment precipitously.

A similar but more complex noncrisis negotiation was played out during the same time frame by Ali Salih ʿAbdullah, the president of the Yemen Arab Republic. For several years, Salih adroitly managed a delicate balancing act between the two superpowers and their respective regional allies, Saudi Arabia and the PDRY. Following the March YAR/PDRY border skirmish, a crisis period already discussed, Salih attempted to forge a close bilateral relationship with the United States along a type VI model; that is, with high U.S. commitment and modest YAR investment. But, in the noncrisis situation that followed the March period, the United States had far more pressing regional and global concerns. The United States, at this point, wanted to emphasize the Saudi-YAR connection. Salih was in no position to pay for U.S. weapons, had no really attractive facilities to offer, and had effectively alienated Moscow to the extent that the United States saw no reason to press ahead with the relationship.[29] Moreover, Saudi Arabia had made it clear to Washington that it considered the Arabian Peninsula its suzerainty and that, although United States help was welcome, it should be channeled through, and administered by, Saudi Arabia.

But, for Salih, a relationship on this basis was unacceptable. Saudi intrigue along the Saudi-YAR border and Riyadh's meddling in internal Yemeni affairs made Salih highly and properly suspicious of Saudi intentions. Salih did not consider the YAR, with the largest national population on the Peninsula, to be Saudi dominions and was not prepared to embark on a relationship that would reinforce this position.

Salih recognized that he must force the United States into a weaker negotiating position. To accomplish this, he began a careful program of overtures to the Soviet Union. Unlike Somalia's unconvincing rapprochement, which was not based on any substance, Salih's efforts had real credibility. The YAR had maintained a military supply relationship with the Soviet Union, even as North Yemeni troops were battling those of the principal Soviet ally on the Peninsula, the PDRY. Soviet advisers had remained in Sana and provided a ready conduit for Salih to begin talking to Moscow in earnest again. Therefore, when Salih sent a military delegation to the Soviet Union to negotiate an arms deal in September 1979, Saudi Arabia and the United States both took the threat seriously.[30]

The Soviet Union, for its part, recognized the opportunity to extend its influence on the Arabian Peninsula and offered Salih liberal repayment terms, fast deliveries, and no Saudi intervention in the arrangement.[31]

The signing of an arms agreement in December 1979, and the arrival in Sana of additional Soviet advisers elicited a U.S. response. In conjunction with Saudi Arabia, the United States applied a variety of pressure techniques, accelerated the deliveries of its own promised equipment, and encouraged Saudi Arabia to be more supportive of Salih. Saudi Arabia, for its part, promised a windfall of economic assistance as well as advisers from Islamic states on the understanding that Salih would reduce the Soviet presence.[32]

Salih accepted all of this in principle and made vague promises to Saudi Arabia and to the United States. But the deliveries of Soviet equipment continued, and Soviet advisers remained in place in the YAR. Salih did concede one critical point to Saudi Arabia and to the United States; the unity talks between Aden and Sana that had grown out of the March conflict were allowed to die, at least for the moment. But the United States and Saudi Arabia feared that a unified Yemen would quickly fall under the domination of the south and, by ready implication, of the Soviet Union.

Because Salih did not face an immediate crisis during this period, the negotiations between the YAR and the United States and, indeed, between the YAR and the Soviet Union, were slow in developing. The nature of the relationship between Sana and Washington remained ill defined, and the nature of the flow of influence was difficult to discern.

As is implicit in the above examples, patron-client state relationships and the negotiations that give them form in noncrisis situations, take place against a background of global issues and are often subordinated to other patron concerns. Patron-client relationships in a noncrisis environment may conveniently be placed on a policy back burner or may be sacrificed by either patron or client in order to serve a larger political objective. Two examples of this emerged during the early months of the Carter administration.

The first of these was the Conventional Arms Transfer (CAT) Talks between the United States and the Soviet Union. These talks were begun with the lofty objective of restraining the sale of weapons to the Third World, under the assumption that this would somehow lower the level of violence. In addition to the arguments that can be made against the wisdom of such an agreement, it is clear that both superpowers were negotiating on a bilateral basis the perceived security of Third World clients. Many states in the Third World see themselves facing high-threat environments, and usually their sole reason for developing ties with patrons is to secure the weapons they feel are necessary for their own defense. Since arms transfers are a principal currency of patron-client state relationships, an agreement to reduce such transfers amounts to an agreement to reduce the relationships they support.

The United States proposed such talks as an almost desperate afterthought following the Soviet rejection of the new U.S. SALT II proposals. Cyrus Vance, seeking urgently to redeem the failure of his March 1977 mission to Moscow, raised the CAT proposal simply to have something to bring back to Washington. The Soviet Union, recognizing that it needed to establish a positive note with the new administration, agreed despite the fact that it, more than the United States, depended upon arms transfers to maintain its position in the Third World.[33] It is highly unlikely that the Soviet Union or even the more sober minds in the United States, ever felt that anything substantial would come of the CAT talks. Indeed, throughout the four rounds of negotiations that were conducted between March 1977 and December 1978, both sides seemed to go out of their way to insure that no progress was made. However, in a world in which perceptions are often more important than reality, the mere fact that the two superpowers were engaged in such negotiations damaged their status with their clients. This shows the willingness of the superpowers to suffer damage in their patron-client state relationships for the sake of only marginal gains in bilateral ties with each other.[34]

In addition to the CAT talks, the United States and the Soviet Union were also engaged in the Indian Ocean Arms Limitation negotiations, ostensibly designed to limit superpower competition in the Indian Ocean. These negotiations, too, were products of the failure of Vance to obtain Soviet agreement to a new SALT II proposal. Although Indian Ocean littoral states were long supportive of the announced underlying objectives of the talks, the specifics that were being discussed appeared to represent an abandonment by the United States of its security responsibilities for the region. Again, in the calculations of the United States and, to a lesser extent, the Soviet Union, progress in bilateral relations with each other was worth the sacrifice of individual relations with Third World clients.

Such subordination of patron-client state relationships is characteristic of noncrisis negotiations. Both patron and client, feeling no particular urgency to press ahead with the relationship, have more time to reflect on the role of the relationship plays in global or regional affairs. Under these conditions, patron-client relationships often become of secondary importance, particularly to the patron.

Crisis Negotiations

The key mechanism for change in patron-client state relationships is the crisis. Crises, by their very nature, dispel the shrouds that cloak the extent of commitment of both patron and client. Moreover, they are the primary vehicles for effecting dramatic change in a relationship within a brief period of time. Crises allow and demand clear statements from both patron and client on the value each places on the other. Crises short-circuit the negotiating process and greatly accelerate the bidding patterns. Significant expansions of commitments are often direct results of crises and, alternatively, the demise of relationships are usually tied to crises as well. Crises, then, set and define the precise nature of a patron-client state relationship at a given point in history.

The power of a crisis to shape the relationship is a function of the characteristics of crisis decision making and, particularly, of the nonrational features that mark the crisis environment. A crisis is an event that comes as a surprise to decision makers, is perceived as a significant threat to highly valued goals, and must be addressed within a short period of time.[35] Crises are not wholly objective phenomena. They are rooted in the decision makers' perceptions of the factors mentioned above. Crisis decision making is characterized by a high sense of urgency, incomplete information, and incrementalism. Grand strategy, or even coherent political tactics are not generally formulated during crises. Policy decisions lurch from issue to issue, with little conceptual regard for overall strategic purpose, except in the most general sense. These characteristics of crisis decision making are rooted in several human and bureaucratic factors.

In any bureaucratic structure, whether it is the Politburo of the CPSU or the National Security Council of the United States, the principal decision maker rarely originates policy options. Rather, he chooses among those presented to him by a small circle of trusted advisers. But even this small circle must rely, in turn, on a group of supporting staff officers, from whom policy options emerge. In a crisis environment, a bureaucratic overload often occurs. The decisions that are made in a crisis are derived from papers written by harried staff officers, usually late at night. These papers cannot and do not address the range of options, nor can they adequately evaluate the flow of intelligence that permeates the crisis environment.[36] So the decision maker must decide among incomplete, often superficial options. In many cases, these decisions are based not upon thorough, persuasive staff work, but rather upon the instincts and deep-seated political philosophies of the decision makers themselves.

Under these conditions, decisions are almost invariably incremental in nature. No decision maker, or member of his supporting staff, wants to adopt a precipitous course of action based upon what he realizes is incomplete or unevaluated information. The overwhelming tendency is to make decisions that allow the government to hold the line and to keep other options open. Each component issue of a crisis becomes a separate entity, not conceptually related to its antecedents or successors. The result is that the decisions, although perhaps individually reasonable and rational, can add up to a major change in policy, one that is probably wholly unintended.

An offshoot of this is that the resolution of a crisis often assumes a dominant role, sometimes to the exclusion of other, objectivity more important, issues. Decision makers and their staffs are limited in their abilities to focus on numbers of issues, and crises tend to rise to levels of importance that may be unwarranted. In the early days of the hostage crisis with Iran, for example, the Special Coordination Committee of the National Security Council (the NSC minus the President) met for several hours each day, focusing the attention of the principal decision makers in the U.S. government and their immediate staffs on that single issue, to the exclusion of issues of more central importance. It has been argued that the U.S. government was unable to respond to the mounting indicators of a Soviet move into Afghanistan because of the fixation on the hostage situation in Iran.[37] Indeed, it can also be argued that the ineffectiveness of the U.S. response over the first several months of the Soviet intervention can be attributed to the unwarranted importance that the hostage situation assumed.

Crises tend to become isolated from the context of other global and regional issues to the extent that options selected for resolution of the crisis may not serve more general strategic purposes. Again, the overload that a crisis often creates prevents the decision makers from taking the time to place the crisis in perspective. The resolution of the crisis becomes paramount in the eyes of the decision makers and this can result in overreaction. This was evident in the decisions made by the United States during the Yemen border skirmish in March 1979. Despite the evidence that the conflict between the two Yemens was a very small affair, of no threat to the security of Saudi Arabia, and did not involve the Soviet Union or its surrogates, the United States reacted to the crisis by ordering the carrier *Constellation* into the Arabian Sea and by shipping large quantities of arms, which exceeded the absorptive capacity of the YAR. Although the overreaction of the United States to the crisis was later cleverly parlayed into a major political advantage, the initial decisions were based, not upon some Machiavellian analysis of the opportunities the crisis presented, but rather upon an intense desire to resolve the crisis on favorable terms, without regard to other consequences.

The implications of these features of crises for patron-client state relationship are profound. The unstable crisis environment, coupled with the inherently unstable patron-client relationship, lead to potententially cataclysmic consequences for the entire international system. The escalation and expansion of a crisis, especially those crises that involve clients of both superpower patrons, present very real dangers that extend far beyond the immediate confines of the relationship itself.

At the outset of any discussion of the role of crises in shaping patron-client state relationships, it is important to draw an analytical distinction between crises and periods of high threat. Although high threat is an essential ingredient in any crisis, the two are not synonymous; high threat is a necessary but not sufficient condition for a crisis. It is entirely possible to have a state in a period of high threat that is not a crisis; Qaddafhi's Libya, for example, existed in an almost perpetual state of high threat, real and imagined, but experienced only one or two crises during the entire decade of the 1970s. This means that relationships of types I, II, and III are not necessarily crisis dependent. Therefore, the client state cannot be expected to exhibit crisislike behavior simply because it faces a high-threat environment. It is the surprise inherent in a crisis and the pressure felt by decisionmakers that result in the all-consuming attention paid to crises and that make the difference in shaping the basic patron-client relationship.

Crises involving the client are more significant in shaping patron-client state relationships and occur more frequently than those that involve the patron. This is because, in client crises, the relationship of the client to its patron often becomes the sine qua non for successful resolution of the crisis. In crises primarily involving the patron, its relationship with the client will probably become but one factor in an array of considerations.

However, the origin of a crisis, in terms of shaping the relationship, is less important than the nature of the crisis itself. A crisis originating with either the patron or the client may quickly become a crisis for both.

From the client's perspective, a crisis can often involve its very survival. Client states generally lack the global economic, political, and military power to absorb major reversals. This is particularly true since most client regimes do not have the institutional depth to weather major political or security upheavals.

Therefore, the client will feel it desperately needs its patron's support and will, in many instances, offer extensive concessions to the patron in exchange. Under these circumstances, the client can be expected to offer such concessions as base rights, treaties, internal realignments, or drastic changes in its international associations. Concessions that the client had been unwilling to make prior to the crisis, may well be forthcoming in the face of patron demands and the exigencies of the crisis. Ethiopia, for example, had resisted Soviet pressure for access to facilities along the Red Sea coast. But, Addis Ababa, when faced with simultaneous crises in Eritrea and in the Ogaden, acceded to Soviet demands and extended Masawa and Assab to Moscow.[38] Similarly, the Soviet Union achieved certain privileges at Cam Ranh Bay in Vietnam as a result of Soviet support during the Vietnam-China border conflict.[39]

But a crisis to the client is a two-edged sword for the patron. It not only offers opportunities; it also entails liabilities. The client, urgently demanding immediate support, imposes the crisis upon the patron.

A crisis involving the client compels the patron to make an immediate and thorough reevaluation of its role and goals, and the level of support it is willing to render in order to sustain the relationship. This reevaluation comes under conditions of high pressure, often forcing the discarding of carefully worked out plans for the client and the relationship. The deeper the crisis becomes for the client, the more substantial are the demands placed on the patron for support. Having rendered some degree of increased support to the client during the initial stages of the crisis, the patron may become involved in an escalation of support, should the crisis to the client deepen. Incrementalism, which marks crisis decision-making, can lead the patron to exceed by far its original intentions in support of the client.

This was the case in the 1979 Soviet-Afghan situation that led to the Soviet intervention. After the ouster and execution of Taraki in September, the insurgency against the Khalq expanded significantly until, by December, it had reached crisis proportions. The Soviet commitment to the relationship, then in the form of large numbers of advisers and economic support, did nothing to stem the insurgency. Indeed, Soviet troops themselves became popular targets for insurgent attacks, often with the most grisly results.[40]

By December, Afghanistan was no longer simply an irritant along a reasonably unimportant frontier of the Soviet Union. It had become, in Soviet eyes, a major crisis that could spread into the contiguous Muslim republics of the Soviet Union and provide the United States with an opportunity to swing Afghanistan into the Western camp.

To the regime of Hafizullah Amin, the crisis had reached unmanageable proportions. Although details of that period remain unclear, it is likely that Amin asked Moscow for more substantial support, even in the form of Soviet combat units in order to suppress the insurgency. The Soviet Union, having committed itself with increasingly large numbers of advisers, responded by sending five divisions across the border, by having Amin executed, and by bringing Babrak Karmal, a hand-picked Parchamist successor, to power in Kabul.

The Soviet commitment of its armed forces represented a major departure from its previous policies; it was the first time that Moscow had sent the Soviet Army outside of the Warsaw Pact countries in order to redeem a political failure. The response of the international community to Soviet aggression was substantial. The loss of the SALT II treaty, the international economic boycott, the grain embargo, the increased restrictions on the transfer of high technology, and even the boycott of the Moscow Olympics were all in direct reprisal. Moreover, the Soviet reputation in the Third World suffered possibly irreparable damage.

In retrospect, the Soviet decision to send its forces into Afghanistan is reflective of the features of crisis decision making discussed earlier. As the crisis to the Khalq regime mounted, the Soviet commitment increased. The movement of Soviet forces, although a major departure from past practices, was an incremental step brought about by Moscow's inability to extract itself from the crisis and by the unwillingness of decision makers in the Kremlin to place the Afghanistan situation, and all the ramifications of Soviet intervention, into a more global political perspective. This case illustrates that, in a crisis environment, decisions are made and commitments are undertaken that, in the cold light of rationality and under noncrisis conditions, might have been avoided.

The Soviet-Afghan crisis also illustrates one of the major dangers inherent in patron-client state relationships, especially those of type III. The Soviet Union, over the course of the four months between the assassination of Taraki and the Soviet intervention, surrendered much of the control over its options to the actions of the client. Hafizullah Amin, the most brutal and doctrinaire of the Khalq leadership, only exacerbated the problems with the insurgency by alienating large segments of the Afghan population, including the military, by his socialist-oriented actions. The Soviet Union was unable to control Amin's activities, even while it was pouring personnel and money into the situation. The decision to commit Soviet troops was brought about not by deliberate Soviet design but through a combination of Moscow's strategic advantage orientation in the relationship and irresponsible Afghani behavior.

Patron-client state relationships carry with them this very real possibility that the patron may lose control over its resources and over its political options, surrendering them to the behavior of often irresponsible client states. This, in turn, can bring patron states into the very sorts of confrontation that the entire patron-client state relationship system is largely designed to avoid.

A crisis, then, is the principal mechanism for radical change in patron-client state relationships. Rarely are relationships unchanged after a crisis. Either the patron and the client have responded to each other's needs, in which case the relationship will continue on a higher level of commitment, or either the patron or the client has not met the other's requirements, in which case the relationship will wither and die.

The tendency is for the former to occur. It is difficult for any patron or client to abandon the relationship during a crisis and in its aftermath. The patron has invested an often substantial amount in the relationship; the client leadership often has become personally identified with the patron to the extent that disengagement is infeasible. Moreover, because of the incrementalism that marks crisis decision making, political momentum often builds up during a crisis, making disengagement more difficult.

In addition to shaping the relationship, crises also clarify the extent of the patron's commitment. This is because the actions required of the patron are frequently visible and widely publicized by the patron, the client, or the adversaries of either. Crises, as attention focusing events, tend to expose actions that under other circumstances might go unnoticed. The massive airlift that the Soviet Union used to resupply Ethiopia in 1977–78, the dispatch of the *Constellation* battle group during the Yemen conflict, or the presence of 85,000 Soviet troops in Afghanistan leaves little ambiguity as to the nature of the patron-client state relationship or of the commitment of the patron. Similarly, the refusal of the United States to send naval forces to support Somalia in 1977, the abject withdrawal of Soviet missiles from Cuba in 1962, and the unwillingness of the United States to agree to huge amounts of security assistance for Pakistan in 1980 all clearly show the limitations of the patron's commitment to its clients.

Crisis Manipulation

It is implicit in the above discussion that crises offer both liabilities and opportunities. Crises are high-risk, high-payoff situations in which both partners in the relationship can gain or lose much. Played against a background of fundamental incompatibility between patron and client, crises present not only volatile problems for the international community but also manipulative tools for patron or client mischief.

Because of the opportunities that present themselves in crises, the patron or the client may attempt to manipulate crises to its own advantage. This is particularly true in cases where noncrisis negotiations have failed to yield the desired results to either partner. The patron, through manipulation of crises, tries to drive the client's commitment from row 2 to row 1 (see Figure 2.3) while holding the patron's commitment constant. The client, using similar logic, tries to force the patron's commitment towards the right (see Figure 2.3) while holding its commitment constant. Because crisis manipulation is not a crisis in the truest sense of the definition, the client or patron may see it as an only moderately risky option. The "crisis" does not come as a surprise to at least one of the partners in the relationship, and, therefore, the dangers of crisis decision making may not appear to be as relevant.

There are two primary forms of crisis manipulation: crisis exaggeration and crisis fabrication. Each involves the creation of an impression of high, immediate threat for the purposes of influencing the partner in the relationship.

Crisis exaggeration is the less dangerous of the manipulative techniques for it involves a largely verbal intensification of an existing set of circumstances in order to present it in a more dangerous light. It does not usually involve an actual escalation of the danger, although that may occur as an unintended consequence. Under this technique, the patron or the client will take a high-threat situation and try to convince its partner that the threat is actually much greater and more immediate. The manipulator hopes that its partner will become convinced of this and will commit itself to greater concessions than it had been willing to make previously. If the gambit fails, only the credibility of the manipulator will be damaged; there is, in theory, little risk of actual escalation of the crisis. This can be done through a variety of techniques, the most common of which is shared "intelligence." The patron or the client, alleging to be privy to sensitive information, will try to create a heightened sense of fear. The Soviet Union, throughout 1979, tried to convince the Cape Verde Islands that invasion and/or subversion was imminent and that Soviet support was the only saving alternative. The Soviet objective was to extract naval basing or ASW aircraft staging rights astride the sea lines of communication in the mid-Atlantic. In this case, crisis exaggeration failed, although the Cape Verde Islands did experience an upheaval of sorts when its partner on the continent, Guinea-Bissau, experienced a coup in 1980.

The United States has been both manipulated and the manipulator in recent years. Exaggeration of crisis is a time-honored Israeli tradition for both justifying its own actions and extracting larger commitments from the United States. United States-Saudi relations have been marked by crisis exaggeration as well. Saudi Arabia successfully manipulated the Yemen crisis, already discussed, to extract a larger U.S. commitment to regional security on the Arabian Peninsula. The United States, for its part, used the threat posed by the Iran-Iraq war in order to station U.S. AWACS aircraft in Saudi Arabia and to intensify the struggling strategic dialogue between the two states. Oman used the threat to the Strait of Hormuz, never very convincing, to gain larger concessions from the United States and from Western Europe.[41]

In none of these cases was the threat actually as great as presented, and never did any of the partners take concrete action to expand the threat artificially. Had the efforts failed in any of these cases, no objective increase in danger would have occurred.

Such is not the case with the second crisis manipulation technique, crisis fabrication. Under this technique, one partner in the relationship, usually the client, announces its intention to create a crisis situation that both the patron and the client know the client cannot win. It is a technique of desperation, adopted when all other forms of negotiation have failed. As will be discussed at more length in the succeeding chapters, Sadat's decisions in 1971, 1972, and 1973 to renew the war with Israel are among the best examples of this technique. Syngman Rhee threatened, in 1953, to launch his South Korean divisions across the cease-fire lines at the Chinese Army, knowing full well that his forces would be decimated. His intentions were to force a resumption of the conflict and break down the talks, so that a more favorable agreement could result.[42]

Crisis fabrication is a highly dangerous manipulative technique. If the patron does not react in the manner desired, the client may find itself either having to back down or to engage in an artificially created situation that could threaten its very existence.

Crisis fabrication, if successful from the client's perspective, has great dangers for the international system. Under conditions of crisis fabrication, the patron stands to lose the most control over its resources and political options. The client, having planned the "crisis," is in a far better position to act rationally during its execution. To the patron, if it accepts the ruse, the crisis is genuine, and all the problems and dangers inherent in crisis management become operative. Concessions and commitments that it might otherwise not make, may be made during the artificial crisis, which, of course, is exactly what the client seeks.

A manipulated crisis, then, is a political and military vehicle by which the client attempts to overload the influence distribution and assume some measure of control over the patron's freedom of choice. When this occurs, the client state, unused to the awesome responsibilities of nuclear confrontation and lacking any real experience in crisis management, may involve the superpowers in the very sorts of conflict that they most want to avoid.

Crises, whether genuine or contrived, are the principal causes of major change in patron-client state relationships over a brief period of time. The breakdown in rational, or at least preplanned, decision making that occurs in crisis situations causes this sort of dramatic change and carries with it significant implications for the entire international community.

CONCLUSION

Patron-client state relationships are rooted in change that derives from the basic incongruity of interests that form their foundation. As internal and external factors arise that put pressure on these relationships, their inherently unstable nature leads to radical, and sometimes rapid, change. The mechanisms by which these relationships change provide ample opportunity for international mischief and, ultimately, for patron conflict.

The propensity for change differentiates patron-client state relationships from historically based partnerships between equals. In the latter case, change does not easily occur, even in the face of substantial pressure. These relationships are predictable and manageable. The fundamental danger in patron-client state relationships lies in their nature and in their susceptibility to violent, unpredictable, and unmanageable change.

NOTES

1. The Soviet Navy, during this period, more than doubled the size of its combatant fleet and moved from a coastal defense force to a blue-water navy, capable of competing with the United States almost anywhere in the world.

2. Soviet ship-days, a measure of the extent of presence over time, increased sixfold in the Indian Ocean between 1970 and 1980. In addition, the Soviet Union deployed larger combatants, up to the verticle take off and landing (VTOL) carrier *Minsk*, during the latter portion of the decade.

3. Michael McGwire, "The Rationale for the Development of Soviet Seapower," in *U.S. Naval Institute Proceedings*, May 1980, pp. 160–62.

4. Alvin Z. Rubinstein, "The Evolution of Soviet Strategy in the Middle East," *Orbis*, Summer 1980, p. 331. IL-38s were staged regularly out of Ethiopia and the PDRY by the early 1980s.

5. Alexander George and Richard Smoke, *Deterrence in American Foreign Policy: Theory and Practice*, (New York: Columbia University Press, 1974), pp. 457–59.

6. Gabriel Marcella, "The Soviet-Cuban Relationship: Symbiotic or Parasitic?" Paper prepared for the U.S. Army War College Strategic Studies Institute Foreign Policy Symposium, September 1979, Carlyle Barracks, Penn.

7. Cuba's role as chairman of the Non-Aligned Movement from 1979 to 1982 further enhanced its international prestige.

8. John B. Lynch, "The Superpowers' Tug of War Over Yemen," in *Military Review*, March 1981, pp. 19–20.

9. *New York Times*, March 10, 1979, p. 1.

10. Abdul Kasim Mansur, "The Military Balance in the Persian Gulf: Who Will Guard the Gulf States From Their Guardians," *Armed Forces Journal International*, November 1980, p. 65.

11. *New York Times*, December 13, 1978, p. 9; December 14, 1978, p. 14.

12. *New York Times*, April 22, 1980, p. 4. In May 1981, Ali Antar himself was forced to resign his post as defense minister and member of the Politburo.

13. *New York Times*, June, 15, 1980; Amos Perlmutter, "The Yemen Strategy," in *New Republic* June 5 and 12, 1980, pp. 16–17, argues that this was only a charade on the part of Al-Hasani.

14. A more extensive discussion of the foreign policy and decision-making implications of this construction appears in Chapter 6.

15. President Carter, in his 1980 State of the Union Address, said:

> An attempt by any outside force to gain control of the Persian Gulf region will be regarded as an assault on the vital interests of the United States. It will be repelled by the use of any means necessary, including military force.

16. Irridentism marks almost every Somali political action. Even the Somali flag is not immune. It has a five pointed star; each point representing a portion of "Greater Somaliland." Three of the points are Djibouti, the Ogaden, and northeast Kenya.

17. A concise summary of the Carter policy towards the Indian Ocean in the first three years of his administration is contained in George W. Shepherd, "Demilitarization Proposals for the Indian Ocean," in *The Indian Ocean in Global Politics* ed. Larry W. Bowman and Ian Clark (Boulder, Col.: Westview Press, 1981), pp. 238–41.

18. The draft agreement called for each side to maintain only one base in the Indian Ocean region; the U.S. base would be Diego Garcia.

19. This modest U.S. answer caused Somalia to give up in the Ogaden. On March 9, 1978, Siad announced that Somali troops had withdrawn from the

disputed area. See Steven David, "Realignment in the Horn: The Soviet Advantage," in *International Security*, Fall 1979, p. 80.

20. The debate was decided on the basis of the historical experience of sending the *Enterprise* battle group to the Arabian Sea in response to the 1971 India-Pakistan war. The unwillingness of the United States to use the *Enterprise* once it got there damaged U.S. credibility.

21. The team was led by Assistant Secretary of State Richard Moose. *New York Times*, March 17, 1979, p. 7.

22. The U.S. team consisted of Robert J. Murray, deputy assistant secretary in Department of Defense; Reginald Bartholomew, director of political-military affairs, State Department; and Fritz Ermarth, NSC staff.

23. The agreement with Oman was concluded on June 3, 1980. The agreement with Kenya was signed on June 12, 1980.

24. *New York Times*, April 22, 1980, p. 3. By May, Siad's figures had dropped to $200 million, which he also leaked to the press. *New York Times*, May 21, 1980, p. 7.

25. *New York Times*, June 25, 1980, p. 4.

26. In the end, a general statement about U.S. concern for Somali security, with appropriate caveats for the Ogaden, was the solution to Siad's concern for security guarantees.

27. *New York Times*, August 19, 1980, p. 10. The agreement was initialed on August 19 and finally signed on August 22 in Washington during a visit by Somali adviser Samantar.

28. These included Congressional restrictions (the "Long Amendment") that required the president to affirm that no Somali troops were in the Ogaden before spending any military construction funds there. *New York Times*, October 1, 1980, p. 9.

29. Abdul Kasim Mansur, "The Military Balance in the Persian Gulf: Who Will Guard the Gulf States From Their Guardians?" in *Armed Forces Journal International*, November 1980, p. 64.

30. Amos Perlmutter, "The Yemen Strategy," p. 16. The Yemeni team obtained agreement from Moscow to provide Su-22s, MiG-21s, and T-54/55s.

31. Mansur, "The Military Balance in the Persian Gulf," p. 64, *New York Times*, March 19, 1980, p. 11.

32. Hermann F. Eilts, "Security Considerations in the Persian Gulf," in *International Security*, Fall 1980, p. 97. *New York Times*, March 19, 1980, p. 1.

33. *Communist Aid to Less Developed Countries of the Free World*, (Washington: CIA, 1978), pp. 1, 6.

34. Linda P. Brady and Ilan Peleg, "Carter's Policy on the supply of Conventional Weapons: Cultural Origins and Diplomatic Consequences," in *Crossroads*, Winter 1980, pp. 41–60.

35. Charles F. Hermann, "International Crises as a Situation Variable," in *International Politics and Foreign Policy*, ed. James N. Rosenau, (New York: The Free Press, 1969), p. 414.

36. Ironically, it is often the overabundance of intelligence, rather than the dearth of it, that gives the decision maker and his supporting staff the most difficulty. Under crisis conditions, all assets of a national intelligence service are mobilized and focused, with a resulting flow of unevaluated and often contradictory information that is awesome in nature.

37. William E. Griffith, "The Implications of Afghanistan," in *Survival*, July-August 1980, p. 148. Griffith was an NSC consultant at the time.

38. Peter Vanneman and Martin James, "Soviet Intervention in the Horn of Africa: Intentions and Implications," in *Policy Review*, Summer 1978, p. 21.

39. Paul J. Nitze, "Strategy in the Decade of the 1980s," in *Foreign Affairs*, Fall 1980, p. 89.

40. Beheadings, torture, and emasculation of Soviet personnel by Afghan insurgents were common. The insurgents would often use Afghan females as bait to lure the untutored Soviet enlisted personnel into ambushes. Soviet officers were particularly attractive targets for insurgent attack as well.

41. Rhetoric of the time notwithstanding, the Iranian Navy really had no capability to close the Strait of Hormuz for any length of time. The Strait is some 30 miles wide at its narrowest and too deep to permit any but the most sophisticated mining, which was far beyond Iranian capabilities.

42. T. R. Fehrenbach, *This Kind of War*, (New York: MacMillan, 1963), pp. 646-47. He was bought off by the offer of a mutual defense treaty, $200 million in aid, and an expansion of the Korean Army to 20 divisions.

CHAPTER 5

A RELATIONSHIP DETAILED

INTRODUCTION

To this point, we have discussed patron-client state relationships in their general sense with illustrative examples of specific aspects. We now turn our analytical attention to the relationship between the Soviet Union and Egypt between 1967 and 1974 in order to demonstrate the efficacy of the conceptual construct and the richness and complexity of such relationships in general.

In many ways, the Soviet-Egyptian relationship is the archetype of a patron-client state relationship in the nuclear age. In it, we find most of the analytical aspects of such relationships as outlined in preceding chapters. Moreover, the relationship has clearly defined end points: the 1955 Czechoslovakia arms deal and the unilateral abrogation of the Soviet-Egyptian Treaty of Friendship and Cooperation in 1976.

From a practical standpoint, this particular relationship is appealing for another reason: the wealth of published literature. Of unique value are the writings of Mohammed Heikel and Anwar Sadat, which set the psychological tone for the relationship and establish the perspective of the principal participants.

The Heikel and Sadat books, with the host of other books and articles published on the subject, as well as firsthand materials from journals, mean that the historical events of the relationship are fairly well defined and will not elicit, in themselves, much debate. Attention can thus be focused on the meaning of these events, not only for the Soviet-Egyptian relationship but also for patron-client state relationships in general. But the existence of a great volume of literature on the Soviet-Egyptian relationship does not make the task of understanding easy. Detailed recitation of historical events does not imply understanding of the underlying dimensions of the relationship, nor does it provide intellectual guidance for the comprehension of other relationships.

We do not here intend to offer a definitive work on the precise events of the relationship, nor to trace the entire history of the relationship. That is more appropriately left to historical works, many of which are already in print. Rather, what follows is an effort to identify key events that demonstrate the richness and dynamism of the relationship and help steer analysis into productive channels. The focus of this examination will be the six-year period 1967–73, with particular emphasis on the October War, a crisis of major proportions that exhibited all of the dangers inherent in patron-client state relationships. The Soviet-Egyptian relationship reflected virtually all the types of such relationships outlined in earlier chapters, as well as most of the mechanisms of change.

THE JUNE WAR: TYPE III

The relationship between the Soviet Union and Egypt just prior to the 1967 war was one in flux. Although the two states had enjoyed a long, if uneven, relationship since 1955, Soviet goals had focused on ideology and international solidarity. The concomitant relationship types had presented only a marginal threat to the international system.

The mounting interest within the post-Khrushchev leadership in expanding Soviet power projection capabilities, and the expansion of the Soviet Navy that made power projection possible, wrought some fundamental changes in the nature of the Soviet-Egyptian relationship. The focus of Soviet interests during this period became increasingly centered on the acquisition of base rights for Soviet naval vessels and aircraft in Egypt. This, in effect, moved the relationship to the right (see Figure 2.3) and set the stage for intrarelationship actions that were to have substantial effects on the entire international system for the next seven years.

The beginning of 1967 saw a basic change in the goals the Soviet Union sought in the relationship. No longer was the Soviet Union content to seek only goals of international solidarity; goals of strategic advantage had become first priority. In effect, Moscow sought to move the relationship from type V to type III. However, with a relatively low-threat environment perceived by Egyptian President Nasser, such a movement was extremely unlikely. Soviet persistence in its quest for bases and threats to terminate the relationship if such requests were not granted were likely to provide unproductive and could actually end a relationship, which, up to now, Moscow had found most useful. What the Soviet Union needed was a mechanism to move the relationship in a manner it desired, and the rising tension between Israel and its Arab neighbors in the early months of 1967 was the catalyst the Soviet Union was seeking. Through crisis manipulation, the Soviet Union had the opportunity to achieve its basing objectives, objectives that were otherwise probably unattainable. The basic Soviet strategy was simple: force Egypt into a crisis with Israel, a crisis that would, in turn, force larger concessions to the Soviet Union.

In April 1967, an artillery duel across the Golan Heights expanded into a major air engagement, in which Israeli warplanes shot down five Syrian MiG-21s. This had substantial reverberations not only in Damascus and Cairo, but also in Moscow.

Two weeks after the air battle, the Israeli ambassador in Moscow was presented with a stiff note that asserted that

> the Soviet government is in possession of information about Israeli troop concentrations on the Arab-Israeli borders at the present time. These concentrations are assuming a dangerous character, coinciding as they do with the hostile campaign in Israel against Syria.[1]

This was to become the central theme in Soviet crisis manipulation, a theme that was to be developed more vigorously over the course of the succeeding weeks.

At the same time, the Soviet Union adopted a complementary posture designed to goad Nasser into a more active role in the crisis. Late in April, in a commentary sure to sting the Egyptian leader, it was alleged that

> imperialists and reactionary Arab circles are circulating a story, especially in Syria, that Egypt is participating in the common Arab confrontation with Israel only with words. In fact, it is sitting snugly behind the U.N. forces in the Sinai. . . . It is completely evident that the imperialists by these means wish to stir up anti-Egyptian sentiments.[2]

Nassar was particularly sensitive to such charges because of his desire to recapture the influence and leadership in the Arab world he had once enjoyed. Indeed, Egypt's position in the region was far from enviable. Despite the 1966 mutual-defense agreement, Syria remained, in Sadat's words, "bitterly anti-Egyptian." Iraq was hostile and isolated; the monarchies of the Arabian Peninsula and Jordan were experiencing economic well-being and had no desire to rekindle the revolutionary fires that had burned in the previous decade. Moreover, Egypt was providing no deterrent to, or punishment for, Israeli raids into the Gaza Strip, the West Bank, or the Golan Heights. Even the Yemen civil war had proven to be disastrous; Nasser was trying to disengage, much to the delight of Saudi Arabia. Egyptian prestige in the Arab world had declined to a point lower than any since the 1952 revolution.[3]

Nasser could not, therefore, tolerate the assertion that his support for the confrontation with Israel was in name only. By emphasizing this point strongly, the Soviet Union placed great pressure on Egypt to become more actively involved in the mounting crisis. It was crucial, from the Soviet perspective, that Nasser's threat environment be escalated; only then did the Soviet Union stand a chance of acquiring the basing rights it so dearly desired. So, it helped stir up regional tensions, hoping to bring Egypt into a crisis frenzy, with the concomitant desperation for Soviet support. The Soviet Union accomplished this objective through contriving, then exploiting, the fiction of a massing of Israeli forces opposite the Golan Heights.

In early May 1967, Sadat visited Moscow enroute to North Korea and was told that Soviet intelligence had determined that ten Israeli brigades were deployed for an attack on the Golan Heights that was to take place between May 18 and 20.[4] The Soviet Union had passed the same story to Syria and to its embassy in Cairo for direct conveyance to Nasser. Nasser then found himself faced with a situation in which he had to act decisively. He responded in the manner most desired by Moscow. On May 15, Nasser ordered U.N. Emergency Force (UNEF) out of the Sinai and three days later deployed Egyptian forces into the Sinai adjacent to the Israeli border.

This move met with immediate Soviet approval. On May 22, *Pravda* said:

> Israeli ruling circles openly threaten the start of an armed intervention in Syria on the pretext that it is allegedly responsible for terrorist acts in Israel. The Israel Army Command has ordered a partial mobilization of reservists. Israeli troops are concentrating on the Syrian border and are committing acts of provocation on land and in the air. In connection with the present situation, the United Arab Republic has requested the withdrawal of the UN forces from the Egyptian-Israeli border. The UAR troops are ready to aid Syria in the event of aggression on the part of Israel.[5]

At the same time, Moscow further goaded Nasser into action by comparing Egypt unfavorably with Syria in resisting Israeli aggression.[6] The Soviet Union also encouraged other Arab states to pressure Nasser into supporting Syria more vigorously.

Up to this point, the crisis had developed in a manner favorable to the Soviet Union. The ejection of UNEF placed Egyptian forces directly opposite Israel, greatly increasing Egypt's threat environment. Had the crisis cooled at that point, Moscow perceived that it would have been in a far better position to move the relationship to gain its objectives.

However, crisis manipulation is a dangerous game, as the Soviet Union soon discovered. Nasser, having abetted the escalation of the crisis, felt he was in no position now to turn back. On May 22, Egypt closed the Strait of Tiran to Israeli shipping, a move that virtually guaranteed, in Egyptian eyes, a war with Israel.[7] Both the United States and the Soviet Union shared this view, and for the first time, Moscow began to see the crisis getting out of hand.

On the day Egypt closed the Strait, Lyndon Johnson sent Kosygin a note that contained a veiled threat of U.S.-Soviet confrontation in the Middle East. At the same time, Moscow's own assessment of the likely outcome of a war showed the Arab states at a distinct disadvantage.[8] Given these factors, it is doubtful that Moscow really wanted a war to begin. The Soviet Union had achieved its objectives in the ejection of the U.N. Emergency Force and the juxtaposition of Egyptian and Israeli forces, with the concomitant increase in Nasser's threat environment. There was little advantage to the Soviet Union in pushing the crisis into war, a war that would certainly bring disgrace upon its Arab clients and a possible confrontation with the United States in which the Soviet Union would be at a strategic disadvantage.

The Soviet Union, faced with this predicament, began to try to cool the crisis. In late May, Kosygin told visiting Egyptian Minister of Defense Badran that "we will back you, but you have gained your point. You have won a political victory. Now is the time for compromise."[9] *Pravda* picked up the theme, praising Egypt for its "sober approach" and "restrained tone" to Israeli provocations. *New Times* warned that "the more tension there is, the easier it will be to weaken the progressive Arab states and, at some opportune moment, overthrow their governments."[10] Nasser felt constrained to ignore these messages, announcing instead that "the USSR supports us in this battle and will not allow any power to intervene until matters are restored to what they were in 1956."[11]

Nasser also put the Soviet Union on notice that, its soothing rhetoric notwithstanding, "everything that now happens follows from the information and advice we have received from [the Soviet] government. You are responsible to me for this."[12]

Nasser felt himself to be in a reasonably advantageous position vis-à-vis Israel. Reassured by his chief of staff, that "everything is in tip-top shape," and that he had at least until June 14 to prepare, Nasser felt under no great pressure to step away from the crisis. Moreover, because of information received through a variety of sources, including Soviet Minister of Defense Grechko, Nasser felt that the Soviet Union was firmly behind him, should the United States intervene on Israel's behalf.[13] Confident that he could acquit himself well against Israel in a war and that he was secure from American intervention, Nasser saw his chance to reassert his leadership in the Arab world.

The events of June 5 brought Nasser's optimism and that of his Soviet patron, to an abrupt end. No effort will be made to recount the tactical details of the battle; these are well known. It is suffcient to say only that it was a defeat of historic proportions, in terms of leadership, tactics, and military skill.

Fortunately for the Soviet Union, the conflict was brief. Even Nasser, in the course of but six days, could not expect his Soviet patron to come to his defense and reverse the tactical rout. Had the war gone on for much longer, the Soviet Union would have found itself having to put substance behind its pledges to support Egypt in battle and could have run into a confrontation with the United States over the war.

The defeat of the Arab armies in the Six Day War was a major tactical setback for the Soviet Union and Egypt. Yet, in defeat, the nature of the relationship changed in precisely the manner desired by the Soviet Union when it began to manipulate the crisis early in the year. The irony of the conflict was that, had the Soviet Union succeeded in defusing the crisis it had so carefully manipulated, and had Nasser secured a major political victory without a war, the Soviet Union would probably not have secured the changed relationship it desired. It is doubtful that, even with the Egyptian army facing Israel, Nasser would have perceived his threat environment as sufficiently high to warrant a change in the relationship and to allow the Soviet Union base rights in Egypt. It was only when Nasser surveyed his shattered armies, his occupied territories, and his burned bridges with the West, that he turned completely to the Soviet Union in a type III relationship.

For the Soviet Union, the cataclysmic defeat of Egypt and the rapid end of the war provided it with the responsibility and the opportunity to demonstrate its massive support for Egypt without risking a confrontation with the United States. The need to rebuild

Egypt's armed forces would provide the ideal chance to press for base access. Moreover, the conflict ended any chance of Egyptian rapprochement with the United States. Nasser's claim that the Sixth Fleet aircraft had destroyed the Egyptian Air Force, a claim that was not believed by Nasser himself or even acknowledged by Moscow, and his subsequent severing of diplomatic relations with the United States, eliminated any residual influence the U.S. might have had in Cairo.[14]

The Six Day War, then, marked the shift in the relationship from type V to type III. It also demonstrated the dangers of crisis manipulation. The Soviet Union, clearly and consciously, saw the mounting tensions in early 1967 as a chance to change its relationship with Egypt in the manner it desired. Moscow helped whip up Arab emotions, playing on Nasser's ego and his desire to assert his leadership in the Arab world. With the ejection of UNEF and the direct confrontation of Egypt and Israel, the Soviet Union felt its objectives had been met. But Moscow then found that it could not control the forces that it had helped set loose and found itself carried along into a major crisis, with all the concomitant uncertainties and dangers of escalation. It was, ironically, Israeli military dominance and tactical serendipity that produced an outcome strategically favorable to the Soviet Union.

The war established the framework for the type III relationship, a framework solidified by the late June visit to Cairo by Soviet president Podgorny. From this visit emerged a new, far more intimate relationship, involving substantial commitments from both sides. The Soviet Union made it clear that it was prepared to underwrite the rebuilding of the Egyptian armed forces. Moreover, by deploying extensive numbers of its own advisers, it closely associated itself with the future preparedness and capabilities of the Egyptian armed forces. Egypt, for its part, isolated itself almost completely from all extraregional powers except the Soviet Union and granted Moscow its most coveted prize: bases in Egypt.

But underlying this close association remained the basic incompatibility that plagues all patron-client state relationships. The Soviet Union supported Egypt in order to extend its capabilities to engage the United States. Egypt sought Soviet support because of its need to redress its humiliation at the hands of Israel. As long as Soviet and Egyptian actions served both purposes, the relationship was harmonious. But, when actions served only one or the other, friction was bound to develop.

The difference in patron and client attitudes was reflected in the perspectives of both states on a resumption of the war with Israel. In the Soviet view, the issue should now be placed in the political arena where it was guaranteed to languish. For it was in the Soviet interest to maintain a situation of high tension without allowing Egypt to engage in actual hostilities. Sadat says:

> It was not part of the Soviet leader's plan to have Nasser fight another war; they supplied him with weapons as a courtesy gesture in recognition of his anti-Americn and anti-imperialist stand and in an attempt to secure Soviet presence in the region.[15]

This basic incompatibility was the primary source of patron-client friction in the interwar years.

THE WAR OF ATTRITION

By 1968 it was evident to Nasser that the Soviet Union was perfectly willing to allow the Arab-Israeli conflict to continue unresolved. But Nasser recognized the futility of this diplomatic approach; standing everywhere on Arab lands and increasingly closely allied with the United States, Israel was under no pressure to compromise. Egypt's own unbending opposition to recognition of Israel resonated in Israel's fortress mentality and further served to harden Israeli unwillingness to effect any movement on the political front.

Nasser decided, then, to force the issue with the Soviet Union by escalating his threat environment. Beginning with long-range artillery, he began a gradually increasing confrontation with Israel. The Israeli Air Force soon struck back at Egyptian targets that their artillery could not reach. When these air strikes failed to dissuade Nasser, the Israeli Air Force went after economic targets.

Nasser's provocation of Israel, however, failed to elicit additional support from Moscow. Nasser then decided to increase the level of conflict. With no Soviet concessions, Nasser declared that Egyptian had embarked on a "war of attrition that could last for 100 years."[16] In response, Israel adopted a new strategy, and extended its aerial operations deep into the Egyptian interior. The principal objectives of this air offensive were to destroy the Egyptian air defense capability, reduce the artillery emplacements near the Canal, and to hold the Egyptian population hostage until Nasser called off his war of attrition. So successful was the Israeli operation that, by early November, 1969, Israeli jets owned the skies over Egypt and attacked targets at will.[17]

Nasser, faced with a declining strategic posture of his own creation, found the Soviet Union still unwilling to escalate its involvement. However, the Israeli commando raid on Port Said in which a sensitive Soviet radar was captured, and the air attack on Aly Zabal in which nearly one hundred Egyptian workers were killed, changed Soviet attitudes.

The Soviet Union realized, at this point, that the war of attrition was truly getting out of hand. It was evident that, without substantial Soviet assistance, Egypt would suffer more unanswerable air strikes by Israel. It was also obvious that Nasser had no intention of calling for a cease-fire while he was at such a demonstrated position of strategic and tactical disadvantage. The Soviet leadership assessed that Nasser would not survive a continuation of the status quo. Their choices were, therefore, either to expand greatly their commitment to Egypt, or be prepared to have Nasser fall and lose their greatly treasured access to Egyptian naval facilities.

Faced with such a choice, Moscow chose to acquiesce to Egyptian requirements. Thus, when Nasser traveled to Moscow in late January 1970, his pleas fell on receptive ears. Kosygin announced that the Soviet Union would provide Egypt with SA-3 missiles to provide low-altitude air defense coverage and would augment the SA-2 high-altitude missiles already in the Egyptian inventory. Moscow also agreed to assume responsibility for Egypt's air defenses over the central portions of the country, an unprecedented step. The air defense responsibilities along the Canal would remain in Egyptian hands, at least for the present. The magnitude of this commitment by the Soviet Union was significant; it represented the first time that the Soviet Union had deployed combat forces to an area outside of the Warsaw Pact countries, with the single exception of Cuba, and was the first deployment of Soviet aircraft to the Middle East.

The combination of Soviet-manned SA-3 batteries, Egyptian-manned SA-2 missiles, and Soviet MiG-21s proved sufficient to protect the Egyptian interior. In addition, Soviet assistance allowed Egypt to upgrade its air defense capabilities along the Suez Canal front. The gradual addition of SA-3 low-altitude air defense missiles along the Canal in June and July began to take a toll of Israeli aircraft; during this period, Israel lost five Phantoms and one Skyhawk to SA-3 batteries.

The influx of Soviet air defense weapons and personnel established a parity of sorts along the Suez Canal by July 1970. After more than a year of a generally unsuccessful war of attrition, this situation represented a substantial political victory for Nasser. Sadat said at the time:

> Our task is to counter the Israeli air war. As you observe, we are succeeding. And, on the more general plane, our objective is to create all the conditions for the rectification of the fruits of aggression.[18]

The Egyptian-Soviet tactical success, coupled with the Israeli reluctance to squander its limited air assets, set the conditions for the acceptance of the cease-fire sponsored by the Rogers Initiative and the effective end to the war of attrition.

The war of attrition is an excellent example of successful crisis manipulation. Faced with an unacceptable political and military situation in late 1968, and unable to extract the needed support from his Soviet patron, Nasser launched a war that he knew he could not win. After initial efforts to control the situation failed, Moscow was faced with the pending defeat of its most important regional client. When Nasser added to this his threat to resign and turn the government over to a president who was pro-American,

Moscow found itself with no attractive alternatives to giving in to Nasser's demands. Nasser, thus, precipitated a crisis that he knew he would lose and successfully forced the Soviet Union to accomplish for him the first phase of his three-part strategy: to defend Egypt. Moreover, with the presence of Soviet combat personnel in Egypt, Nasser succeeded in enmeshing his Soviet patron deeper into his war-fighting strategy, making it far easier for Egypt to obtain substantial concessions from the Soviet Union in the future.

THE RELATIONSHIP IN CRISIS: THE DEATH OF NASSER

Gamel ʿAbd el-Nasser's death of a massive heart attack caused justified consternation in Moscow; experience with client leaders such as Nkrumah in Ghana showed that close relations with client states were almost wholly dependent upon the survival in power of individual leaders. Nasser's successor, at least temporarily, the vice president, Anwar Sadat, was an uncertain factor. Indeed, to the Soviet Union, Sadat presented a rather unpleasant prospect. He was renowned for his admiration of the West and, in particular, the United States.[19] He had periodically clashed with Soviet leaders, had displayed violently anticommunist leanings and was a practicing Muslim. In addition, Sadat was not closely associated with Nasser's attempt to assume a dominant and revolutionary political position in the Arab world. This contrasted sharply with Egypt under Nasser's leadership.

Despite periodic clashes, the Soviet Union had, on the whole, been satisfied with its relationship with Nasser. Since 1967 Nasser had given the Soviets access to Egyptian air and naval facilities, the most basic Soviet objective in the relationship. Indeed, by 1970 the Soviet Union was assuming exclusive control over facilities at Alexandria and the newly constructed base at Mersa Metruh. The Soviet Union could not now afford to jeopardize its relationship with Egypt, whatever leadership emerged in the post-Nasser period. Adding to Soviet concern was the visit to Cairo of Richard Nixon's emissary, Elliot Richardson, the first such envoy to be received in

Cairo since relations were severed in 1967. In the uncertain world of Anwar Sadat, the Soviet Union was deeply alarmed by what such a visit could portend.

In this murky environment, Kosygin hastened to Cairo ostensibly to pay homage to the dead Nasser. He was accompanied by a heavy military team, including Soviet air defense specialists. [20] During this visit, Kosygin went to great lengths to announce that Soviet-Eygptian relations were unchanged by the untimely death of Nasser. Indeed, in a joint communique issued upon Kosygin's departure, Sadat and Kosygin pledged to carry out the program of reclaiming the territories lost in the 1967 war and reaffirmed their mutual distaste for imperialisn in the Middle East.[21]

Kosygin realized full well that words can evaporate quickly. He, therefore, provided Sadat with promises of a tangible measure of the Soviet commitment. Moscow offered to increase its support for Egypt and to provide additional military hardware. By the end of the year, substantial quantities of new weapons and Soviet advisers began to appear in Egypt.[22] In private conversations with Egyptian cabinet members, Kosygin also voiced his concern over Egypt's future course. Egypt had to resist "the imperialists [who] were going to attack us now that Nasser was dead."[23]

Nasser could not have picked a more propitious moment to die. As it happened, Nasser had so thoroughly burned Egyptian bridges to the West that a rapprochement was impossible. Sadat points out that "the legacy Nasser left me was in pitiable condition. In the sphere of foreign policy, I found that we had no relations with any country except the Soviet Union."[24]

Sadat was further handicapped by his low standing in the West. Elliot Richardson, far from effecting the rapprochement Moscow feared, reported to Nixon that Sadat would not last for more than a few months.[25] The United States also felt that Sadat was "purely and simply pro-Russian."[26] When he appointed Ali Sabry as his vice president, U.S. fears were seen as confirmed. Sadat thus had the unique distinction of being perceived by each superpower as sympathetic to the other. The one point both Moscow and Washington agreed on, however, was that Sadat would not last.

Within the context of this political isolation, Sadat quickly recognized that his survival depended upon his ability to make progress in regaining the Sinai. This became his overriding national objective. Sadat, at this point, had only the force of arms to turn to, and only the Soviet Union held out any promise of supplying the offensive weapons Egypt needed to carry the war to Israel.

Sadat, therefore, had no options except to maintain Egypt's relationship with the Soviet Union. The endurance of the relationship beyond the death of Nasser was not due to any institutionalization; Sadat would have quickly ended the relationship had Egypt had any other way to address the Israeli problem. The Soviet-Egyptian relationship, which from the Egyptian perspective was created solely by the Israeli occupation of the Sinai and the threat this created, survived Nasser's death for different reasons than its survival after the ouster of Khrushchev in 1964. Krushchev's fall power did not create a crisis in the relationship; changes in patron leadership seldom do, at least in the short term. The death of Nasser, on the other hand, created a genuine crisis in the relationship; the fact that it endured was the result of the immediate situation and Egypt's high-threat environment.

In response to Sadat's initial barrage of pro-Soviet rhetoric, the Soviet Union honored at least a portion of its September promises. By the middle of the fall of 1970, the Eyptian Army boasted over 1,200 tanks. Heavy artillery pieces had doubled over pre-1967 strength, and an additional 50 MiG-21 aircraft had been added to the Egyptian inventory. Missile launchers lined the Suez Canal at intervals of 7.5 miles and covered some 12 miles of airspace over the Israeli-occupied Sinai.[27]

Sadat knew that, despite his election by the National Assembly, his hold on power remained closely wedded to his ability to show movement in the Sinai. As his political options with the West grew increasingly remote, his reliance on the Soviet Union for weapons became the centerpiece of his extraregional foreign policy.

The Soviet Union, for its part, continued to regard Sadat as a caretaker to be courted, but not assiduously. After the first rush of support following the Kosygin visit, the Soviet Union was content to allow events in Cairo to develop on their own. Sadat's initial behavior convinced Moscow that its access to Egyptian naval facilities remained unendangered and that an Egyptian reconciliation with the United States was highly unlikely. The Soviet Union anticipated a coup against Sadat led by Ali Sabry who was, as Moscow well knew, ideologically and pragmatically, "Moscow's man."[28]

However, as long as Sadat did nothing to jeopardize the burgeoning Soviet access to Egyptian facilities, Moscow was prepared to provide some support, albeit unenthusiastically. And, indeed, Sadat did not waver from Nasser's policy of providing facilities for the Soviet Navy. In addition to access to Alexandria and Port Said, the Soviet Union was allowed to exercise complete jurisdiction over the new naval complex at Mersa Metruh. Moreover, Soviet access was not limited to naval facilities. Heikel reports that:

> The Russians virtually took over control of Cairo West airport to the extent that there was not even a representative of the Egyptian customs there. Soviet planes landed and took off as they pleased.[29]

However, the Soviet Union was by no means an open cornucopia of weapons for Sadat. Indeed, the pledges made by Kosygin in September were slow in materializing, giving Sadat his first real exposure to the gap between Soviet promises and Soviet deliveries. By late February 1971, Sadat felt sufficiently frustrated with Soviet intransigence to invite himself to Moscow to confront his patron directly.

Sadat's primary objective in this secret trip to Moscow was to secure firm Soviet agreement on several weapons systems that he felt were indispensable to an effective air defense of upper Egypt, including the much vaunted Aswan High Dam. An effective air defense system was one of the three key military elements Sadat required in order to resume offensive operations against Israel. Sadat realized that Egypt needed an effective air defense to protect its fledging economic infrastructure from Israeli air attacks during a resumed war. In addition, Sadat needed a deterrent weapon and a battlefield air defense system. The March 1971 visit, however, focused on the first objective. In this regard, Sadat asked for additional SA-3 and SA-2 batteries and for the MiG-25 Foxbat, which he felt was the only weapon capable of dealing with the Israeli F-4.[30]

The March meeting was hardly amicable. Instead of responding to Sadat's request, Kosygin and Grechko took turns baiting Sadat and making unreasonable demands of their own. Kosygin offered Sadat the MiG-25 but only with the stipulation that the aircraft would be used only with Soviet permission on a case-by-case basis. MiG-25s were, in fact, already in Egypt, but they were under exclusive control of the Soviet Union and piloted by Soviet airmen. These aircraft were used to shadow the U.S. Sixth Fleet in the Mediterranean. They were also supposed to be available to respond to Egyptian reconnaissance requests, but such requests were systematically turned down by Soviet commanders in Egypt.[31] This gave Sadat an understandably skeptical attitude about any arrangement that allowed the Soviet Union to veto Egyptian military operations. Sadat was, therefore, not pleased by the Soviet suggestion.

Brezhnev, playing the role of peacemaker, intervened with a conciliatory offer of 30 MiG-25s to use as bombers. This did much to placate Sadat, even though there remained some restrictions on the use of the aircraft. Sadat's anger was further damped by the realization that whatever the Soviets were willing to offer was better than nothing. And, indeed, as the meeting wore on, the Soviet Union did offer some additional weapons, primarily air defense batteries. Sadat summed up what happened by saying, "I had to make an angry scene, but in the end, I got what I wanted."[32]

This success nothwithstanding, Sadat returned to Cairo with a bitter and defensive attitude towards the Soviet-Egyptian relationship and a commitment to end his ties with Moscow just as soon as the issue with Israel was settled. He realized that the Soviet Union had no confidence in him and that he was not

> Moscow's man and never would be. They considered me temporary until someone else from the centers of power [Sadat's term for the Ali Sabry faction] took over. They infiltrated positions in the media, the

> press, and in many places in Egypt. Everything was progressing to prepare for the day when [Sabry] would take over.[33]

From the Soviet perspective, the March trip ended the crisis brought on by the death of Nasser. After Sadat left, the Soviet leadership felt that its grip on its client was as firm as ever, perhaps even more so, since they remained convinced that Sadat would soon fall to a Sabry-led coup. In the meantime, Sadat had demonstrated his willingness to come to Soviet terms on arms supply issues and to maintain the Soviet presence in Egypt.

The type III relationship survived the death of Nasser primarily because Sadat had nowhere else to go for the support he needed to reduce his threat environment. Moreover, although the Soviet Union did not meet all of Sadat's requirements, Moscow did prove itself willing to expand its arms supplies to a modest degree.

But the durability of the relationship was illusory; the seeds of its destruction were sown irreversibly by the death of Nasser. Sadat truly despised the Soviet Union and, in a Third World client state, such an attitude held by the leader will shape the future of the relationship. In addition, Sadat was in a better position to achieve a reduction in his threat environment, short of the near impossible task of ejecting Israel from the Sinai with military force. Although Sadat had been a close personal friend of Nasser, he had never associated himself with Nasser's pledges to destroy Israel. Although Nasser abandoned this rhetoric after the 1967 war, his personal commitment to the destruction of Israel remained an important impediment to any negotiated settlement. Sadat, unhindered by this rhetorical baggage, was in a better position to negotiate a peace, once his military bona fides were established in a limited war. Thus, the Soviet grip on Sadat was potentially far less secure than it had been with Nasser because of Sadat's ability to reduce Egypt's threat environment without Moscow's help.

In the interval between March and May 1971, Soviet attitudes toward the relationship had gone from confidence to concern. By late May, it had become apparent in Moscow that Sadat's hold on power was growing stronger and the Soviet strategy of holding Sadat at arm's length until a successor came to power was not bearing fruit. At the same time, Sadat was apparently displaying a willingness to deal with the West (symbolized by the Rogers visit) and a political flexibility that Nasser had lacked in seeking a solution to the Arab-Israeli conflict. Sadat felt that the time was not yet ripe for having the United States assume the role of regional arbiter, but Moscow's perception of the possibilities was less sanguine. Rhetoric notwithstanding, a resolution to the Arab-Israeli conflict was the last thing Moscow sought. An elimination or substantial reduction in Sadat's threat environment, especially if such a reduction were accomplished under the auspices of the United States, would drive the relationship from type III to type VI. Under such circumstances, either the relationship would die (a likely outcome, given Sadat's anti-Soviet views), or Egypt would assume a position of dominance. Both outcomes would be equally damaging to Soviet interests and would spell the end of the substantial Soviet access to Egyptian facilities. The Soviet Union knew that, without a pressing requirement for Soviet weapons, Sadat's latent hatred for Moscow would surface, and this was something that Moscow sought desperately to avoid.

Thus, on May 25 Soviet President Podgorny hastened to Cairo with a Treaty of Friendship and Cooperation, ready for Sadat's immediate signature. The treaty did not come as a complete surprise to Sadat; Sami Sharaf, purged in early May, had negotiated the essential ingredients of such a treaty before he was ousted. Nonetheless, the timing and the urgency both surprised Sadat. Although he approved of the idea of a treaty, he was concerned that the timing was wrong.[34] But Podgorny insisted, asserting that the West was making great political capital over the May ouster of the Ali Sabry group and that the loss of Soviet prestige was intolerable. In his toast on the occasion of the signing of the treaty, Podgorny played to this theme:

> The Treaty . . . signifies a new blow to the plans of international imperialism which is trying in every possible way to drive a wedge into the relations between our countries, to undermine our friendship, and to divide the progressive forces.[35]

Sadat, although somewhat reluctantly, decided to sign the treaty for three principal reasons. First, he recognized that there really was no substance behind the Soviet fears of a rapprochement with the West. The purge of Sabry was much more a personal vendetta than a reflection of any real effort to move away from the Soviet Union or to reduce Soviet influence in Cairo.[36] Second, Podgorny brought with him a firm pledge that all the weapons the Egyptians has asked for, including the retaliation weapon, would be shipped within four or five days.[37] This included the 30 MiG-25s, with no restrictions whatever on their use. Third, the Treaty of Friendship and Cooperation contained an article that committed the Soviet Union to supply weapons and training "to use in the cause of liquidating the consequences of aggression," thus specifically tying Moscow to a military solution to the continued occupation of the Sinai by Israel.[38] For Sadat, this last reason was the most compelling.

To the two sides, the treaty represented entirely different levels of commitment. For the Soviet Union, the treaty was a formal ratification of what it thought it had gained in the March 1971 Sadat visit to Moscow. The treaty, in Podgorny's words, was designed to "strengthen and cement what had matured in Soviet-Egyptian relations in recent years," a maturation process that presumably included Soviet access to Egyptian facilities.[39] Moscow did not intend that this treaty represent a new level in the relationship or that it imply Moscow's acceptance of Sadat's military aspirations in the Sinai. The treaty, in Soviet eyes, was designed to arrest what Moscow saw as a dangerous trend in its relationship with Egypt, a trend that required important countermeasures. Indeed, the Treaty of Friendship and Cooperation was such a measure; it was the first of its kind between the Soviet Union and any state of the Third World. The Soviet Union did not embark on this treaty as casually, nor did the Soviet Union use the treaty solely to demonstrate international solidarity, as it would with future such treaties. The demonstrated international solidarity was but an immediate by-product designed to drive the United States closer to Israel and out of the role of a potentially effective arbiter in the Arab-Israeli dispute. The real Soviet objective remained the preservation of its relationship with Egypt, and more importantly, the continued improvement of Soviet access to Egyptian bases.

Sadat, on the other hand, saw the treaty as Soviet approval to renew the war against Israel and a commitment by the Soviet Union to provide him the weapons he needed for such a conflict. In the treaty, Sadat saw as forthcoming the equipment and guarantees that he felt were required for a successful attack across the Canal to liberate the Sinai. The treaty, thus, to Sadat represented a new level in the relationship.

With this basic incompatibility of views on the meaning of the treaty and the commitments it implied, the benefits to the relationship were predictably ephemeral. But, although the treaty would have no long-term effects on the nature of the type III relationship, it did formalize the relationship and provided a useful point of departure for future discussions.

CRISIS MANIPULATION: THE "YEAR OF DECISION"

Although Sadat had high hopes for the relationship after the signing of the treaty, he was realistic enough to recognize the Soviet propensity for making promises it had no intention of fulfilling. Indeed, after Podgorny's pledge of major arms deliveries within four or five days after the signing of the treaty, no weapons were delivered for three months. Sadat, therefore, decided to maintain the pressure on the Soviet Union by declaring that 1971 was the "year of decision" for the situation in the Middle East. This modest rhetorical effort at crisis manipulation was designed to force Moscow to honor its commitments to provide Sadat with the weapons he needed. Implicit in this announcement was Sadat's threat that he would attack with or without the weapons and would risk a defeat at the hands Israel rather than allow the stalemate to continue. This would, in turn, reflect negatively on Egypt's new treaty partner and recognized patron.

The Soviet Union, unimpressed with Sadat's pronouncement, did respond by providing Egypt with certain amphibious equipment, assault bridges, and other river-crossing materiel. But other weapons were not forthcoming. More than ever, the Soviet Union wanted to avoid a renewal of the war. Having used the treaty to polarize the superpower positions in the Middle East, any conflict would certainly bring about a U.S.-Soviet confrontation as each side sought to support its client. However, at the 24th Congress of the Communist Party of the Soviet Union (CPSU), it was decided that the Soviet Union would pursue a policy of détente with the United States, and some negotiations were already under way between the superpowers to limit strategic arms. The Soviet Union also assumed that, with the treaty, it had regained the upper hand in the relationship and that it could afford to ignore its commitments to Egypt.

The Soviet leadership treated Sadat with a scornful and cavalier attitude during this period, in marked contrast to the care with which they had treated Nasser the year before. When it became apparent to Sadat that his "year of decision" would be military suicide without Soviet weapons and that those weapons were not forthcoming, he appealed to the Soviet Union for support. He reminded them, through the Soviet ambassador, V. Vinagradov, of the promises Podgorny had made in May. However, still no Soviet support was forthcoming.

Sadat's analysis of the needs of the Egyptian army, completed in mid-1971, showed that he required both qualitative and quantitative support. Egypt primarily needed a mobile air defense system that would accompany the advancing forces and neutralize the advantage of the Israeli Air Force all over the battlefield.

In addition, Sadat sought more and better antitank weapons in order to blunt an Israeli armored counterattack. The Sagger antitank guided missile (ATGM) was a particularly attractive system that could be made available in great numbers to the Egyptian Army.

Finally Sadat sought a weapon with which he could threaten Israeli population centers in order to deter an attack by Israeli F-4s on Cairo. This could take the form of either a medium-range bomber, such as the Tu-22, or an intermediate-range rocket, such as the Scud B.

If Sadat were to honor his pledge to make 1971 "year of decision," he would have to acquire all these weapons well before the end of the year in order to allow his armed forces time to absorb the new sytems into their inventories.

With this in mind, Sadat invited himself to Moscow in October to make a final plea. As a gesture to Moscow, and in response to Soviet pressures, he had already appointed Murad Gareb, a man of renowned pro-Soviet leanings, to be Egypt's Foreign Minister. The meaning of this was obvious; Sadat was prepared to make adjustments to Egypt's internal structure in order to demonstrate his continued solidarity with Moscow and in order to offer an olive branch for his anti-Soviet activities of the summer.[40]

Sadat went to Moscow in desperation; he needed offensive weapons for his "year of decision," and only the Soviet Union could provide them. The trip was initially disappointing. The Soviet Union had no intention of supporting Sadat's irresponsible rhetoric and felt that Sadat's effort at crisis manipulation lacked credibility. When it became apparent that the Soviet Union was not prepared to honor its pledges, however, Sadat escalated the stakes:

> I said: . . . I would like to tell you that our men have been trained on [the Soviet-manned] missiles and can operate them instead of your own men. . . . Therefore, it is time to withdraw the Soviet crews. My words struck the three Soviet leaders like a thunderbolt. It was as if I had directed their missiles at them. . . . [Brezhnev] said: This will be a catastrophe for the Soviet Union. I said: I would like to know how this could be a catastrophe for the Soviet Union. He replied: Soviet presence would be in danger.[41]

The Soviet Union saw Sadat's threat as the beginnings of a reduction in Soviet access to facilities in Egypt. Although the withdrawal of the missiles crews would not, in itself, place at risk the access to Egypt for operational Soviet units, it would set a disastrous precedent.

Moscow, therefore, responded to this threat, promising Sadat delivery of all the weapons he desired. Sadat reports that he "went back to Egypt, this time fully confident that the promised weapons would soon be on their way to us."[42] Sadat was confident enough to set in motion operational details to renew the war against Israel.

But Sadat's optimism was premature. The Soviet Union, having soothed Sadat with words, now reverted to its practice of promising much but delivering little. It is likely that, after the initial shock of Sadat's threats wore off, the Soviet Union realized that they were hollow, that the "year of decision" was contrived to force more support, and that Sadat was still unable to renew the war.

Whatever Soviet intentions might have been, the 1971 war between India and Pakistan provided Moscow with a more pressing problem. The war provided a lesson to Sadat of the low esteem in which the Soviet Union held Egypt. Not only did Moscow divert some of the equipment that it had designated for Egypt to resupply India, the Soviet Union actually drew down Egyptian stocks positioned at Cairo West airfield and sent them to India.

Sadat was furious not simply because of the callous treatment afforded him by his Soviet patron, but also because Soviet actions had caused him acute embarrassment in the Arab world. Many Arab states were angry at the fact that the Soviet Union played such a major role in the defeat of the Islamic state of Pakistan and that Egypt, the reemerging leader of the Arab world, was an accomplice of this action.[43]

As a further affront to the Arab world, the Soviet Union modified its emigration regulations in the last part of 1971 to permit a tenfold increase in the number of Soviet Jews allowed to leave the Soviet Union. This action, made as a gesture to the United States, was seen in the Arab world as a direct Soviet contribution to the armed forces of Israel, since most of the Jews allowed to leave the Soviet Union were of military age.[44] Although not a centerpiece of the Soviet-Egyptian relationship, the question of the emigration of Jews was to continue to plague the relationship throughout its duration.

By the end of the year, relations between the Soviet Union and Egypt had again deteriorated. Sadat was unable to make 1971 his "year of decision," due to the failure of the Soviet Union to deliver the weapons he needed. His failure in the manipulated crisis of the "year of decision" severely damaged his credibility, both internationally and domestically. Opposition inside Egypt grew, both to Sadat and to the Soviet-Egyptian relationship. Although Sadat formally attributed the failure of Egypt to make 1971 the "year of decision" to the unforseen circumstances of the India-Pakistan war, this approach had little credibility inside of Egypt.

The ebb and flow of influence in 1971 is a useful demonstration of the changes within a patron-client state relationship that can occur within the parameters of the general relationship type. Although, throughout 1971 the relationship remained basically type III, the flow of influence between Cairo and Moscow changed substantially. As mentioned in chapter 2, type III relationships are of such a volatile nature that changes in influence are likely to occur frequently and dramatically. In the early part of the year, the Soviet Union was clearly in a subordinate position, as most vividly demonstrated by the supplicant role played by Podgorny in his May trip with the Treaty of Friendship and Cooperation. By the end of the year, however, the influence patterns had changed, as best demonstrated by Moscow's callous treatment in Egypt during the India-Pakistan war. The relationship remained type III throughout the period, because the goals sought by both states themselves remained constant. But because type III relationships, by their very nature, do not dictate dominance by either state, there is a great deal of room for maneuver by both sides. This was evident in the 1971 "year of decision."

1972: MOUNTING TYPE III CRISIS

The pivotal year of 1972 began with relations extremely cool. The mounting Soviet presence in Egypt and continued Soviet acquisition of exclusive rights over various air and naval bases were festering sores with the Egyptian military. The Soviet advisers had an extremely low opinion of their Egyptian counterparts and made no effort to conceal their disdain. They considered Egypt to have a "class army," praising the enlisted men while ridiculing the officer corps. Soviet haughtiness, endemic in Soviet relations throughout the Third World, was intensified by the prickly pride of Egyptian Commander-in-Chief, Mohammed Saddek, who regularly reported to Sadat the slights and insults of the Soviet advisers. Saddek became a hero to the rising numbers of Egyptian officers and, indeed, to the Egyptian people, who opposed Soviet presence in Egypt and the extent of influence exercised by the Soviet Union over Sadat.[45]

In addition, the Soviet Union had become increasingly secretive about its operations at Egyptian bases over which it had exclusive control. This extended to the point of a requirement that Saddek and even Sadat have permission to visit a Soviet base over which Egypt still retained nominal title. Again, to the Egyptians, this represented an unacceptable affront to their nationalistic pride; it was seen as a replay of the activities of the British armed forces prior to 1956.

The frustrations of the Egyptian leadership over the failure of the Soviet Union to honor its commitments during the "year of decision," and the increasing arrogance of the Soviet personnel in Egypt, resulted in a series of student demonstrations in early January 1972. Soviet bookstores and other enterprises were smashed, and demands for Soviet weapons were repeated publicly. Sadat, whatever his personal feelings about the Soviet Union, could not allow this domestic violence to go unchecked. He reacted strongly, arresting large numbers of the demonstrators, which served to quiet the streets in Cairo but spawned further opposition to the Soviet Union, and by implication to Sadat, among Egypt's intellectuals.

Thus, Sadat found himself in an unenviable position in early 1972. His failure to execute the "year of decision" had indeed seriously damaged his domestic position; his defense of the Soviet Union at the time elicited strong nationalistic sentiments. Criticized by the Soviet Union for his domestic policies, pressured by his military to reduce Soviet presence, scorned by his own people and other Arabs for his inability to act in 1971 against Israel, Sadat could not long survive a continuation of this situation.

By the same period, the Soviet Union began to realize the difficult straits in which Sadat found himself. Although Moscow was in no way enamored with Sadat, there were no viable alternative leaders on the horizon who were not anti-Soviet. In addition, the domestic disturbances and attitudes within Egypt virtually guaranteed that, should Sadat fall, the Soviet position in Egypt, and access to Egyptian bases, would be seriously jeopardized. Brezhnev, therefore, decided to invite Sadat to Moscow to review how the relationship might be repaired and Sadat's domestic positioned strengthened.

Thus, when Sadat journeyed to Moscow in early February 1972, he found Brezhnev far more accommodating. Brezhnev accepted all the blame for the failure of the Soviet Union to deliver the promised weapons in 1971, ascribing the Soviet failure to "the inevitable paperwork, the necessary red tape, and similar things."[46] He pledged that the situation would be changed and that the Soviet Union would deliver the promised weapons on an expeditious basis. He further promised additional weapons as well, including the short-range FROG-7 surface-to-surface rocket, additional Tu-22 bombers, and the newest T-62 tank. Kosygin, for his part, promised to assume personal responsibility for expediting deliveries of weapons in the future.[47]

However, Sadat was not pleased. First, he knew well now that Soviet promises and Soviet deliveries were two distinct things. Second, he had not obtained Soviet pledges to deliver the weapons he needed most, as outlined above. Although he was pleased to have the Soviet Union acknowledge its failure in 1971, he saw a similar situation developing in 1972.

And, indeed, Sadat's skepticism was well-founded; the Soviet Union had no intention of providing Sadat with the weapons he needed to resume the war. By 1972 the promise of détente with the United States, through which the Soviet Union could achieve its goals perhaps more directly, led the Soviet Union to seek to avoid any confrontations with the West. A resumption of the war, particularly under conditions of Egyptian military inferiority, would have provided a sure confrontation, as Moscow would have been forced to support Egypt against the American client Israel. This was further reinforced by the approaching summit between Nixon and Brezhnev, which promised to be extremely beneficial for the Soviet Union. Therefore, during the February Sadat visit, Moscow sought to strike a delicate balance between satisfying Sadat's desires and preventing a new round of fighting in the Middle East.

Moscow succeeded in the second, but failed in the first. After ten years of unfulfilled promises, Sadat was not prepared to accept half measures. Sadat vented his anger with his Soviet patron in a series of speeches in which he pledged to resign anytime his people lost confidence in him. He also promised to extract Egypt from the squeeze exerted by the superpower tactics and to seek new support from Asia, "with Peking at its center."[48] He also criticized the quality of Soviet arms that were shipped to Egypt and promised to seek to buy arms from Western Europe. Finally, he released all the anti-Soviet demonstrators who had been arrested in January and alllowed them to return to the newly reopened University of Cairo.[49]

When this had no effect on the Soviet Union, Sadat, under strong pressure from Saddek, expelled the chief Soviet adviser in Egypt. The Egyptian commander-in-chief had become so angry with Soviet high-handedness that he would no longer even speak to any Soviet officer.

Given these developments, the Soviet Union decided that it was time again to consult with Sadat. Brezhnev invited Sadat to Moscow for yet another visit in May 1972, just prior to the Brezhnev-Nixon summit. Soviet objectives in this meeting were twofold. First, Brezhnev wanted to arrest the erosion of the relationship, hopefully with no more than some additional, powerful rhetoric. Second, and in Sadat's eyes, more important, the Soviet Union wanted to demonstrate its control over its Egyptian client as an object lesson for the United States prior to the summit. Sadat related bitterly that the Soviet Union wanted to show the United States the extent of Moscow's influence in the region.

> They [the Soviet leadership] wanted me to pay a visit to Moscow toward the end of April. I realized, of course, that they simply wanted to prove to the United States that their feet were planted firmly in the Middle East region. But despite being disgusted with the whole gesture, I accepted their invitation.[50]

Sadat felt that, whatever the charade might be, he should sieze upon any opportunity to voice his views on the region to his Soviet patron.

Despite Sadat's cynicism, however, he did succeed in obtaining a cleverly crafted Soviet commitment to provide the necessary offensive arms if Moscow could not convince the United States to pressure Israel into an acceptable political solution.[51] Moscow agreed that a five-month deadline for political movement was appropriate and that, should no progress be made within this time frame, the Soviet Union would support an Egyptian military solution. The joint communique issued as Sadat left Moscow hinted at this agreement. It contained a line saying that the Arabs had the "right to use other [than political] means," to recover territory occupied by Israel.[52] This represented a significant departure from the rigid Soviet adherence to rhetoric supporting only a peaceful solution to the Middle East issue which had marked Soviet communiques in the past.

The Soviet Union also agreed to provide Sadat with a detailed analysis of the Nixon-Brezhnev summit immediately upon its conclusion. Brezhnev stressed to Sadat that the Soviet Union was not going to sacrifice Egyptian interests simply to move détente with the United States farther along.

His February and May visits to the Soviet Union, however, did little to quell the rising anti-Sovietism in Egypt or to solidify Sadat's own domestic position. Virtually on a weekly basis, there were additional manifestations of the deep-seated and growing discontent within Egypt with the "no war, no peace" situation. In addition, Sadat received further indications of serious antipathy within the officer corps of the Egyptian armed forces on whom Sadat ultimately depended for his political security. A formal petition was submitted to Sadat by the officer corps urging him to reduce the Soviet presence in Egypt and to regain control of Egyptian military facilities. Sadat had no viable response to the powerful and restive Egyptian military because he could show no substantial increase in Soviet weaponry, nor could he promise an immediate resumption of hostilities against Israel.

As far as weapons deliveries were concerned, none was forthcoming. The April agreement with Moscow had put off the needed offensive weapons until November at the earliest. Soviet Defense Minister, Marshal A. Grechko, however, did travel to Cairo shortly before the Nixon-Brezhnev summit to show Sadat the new version of the MiG-25 *Foxbat*.[53] The high-flying MiG-25 made several highly publicized overflights of the Israeli-occupied Sinai to demonstrate the superiority of Soviet technology over the U.S.-supplied F-4 *Phantom*. Grechko's visit was also designed to underscore Soviet influence in the Middle East on the eve of the Nixon-Brezhnev summit. In this regard, Sadat again says that the Soviet purpose in the visit was to "demonstrate Soviet influence in the Middle East," five days before Nixon arrived in Moscow for the summit. Sadat did not mind, however, because he felt that Grechko's visit might portend a more serious Soviet attitude towards Egyptian military needs.[54]

The inclusion of the MiG-25 in Grechko's party was a curious affair, for the Soviets took the aircraft back with them when they departed Cairo. Grechko may have attempted to barter the MiG-25 for more extensive base rights at Ras Banas (Berenice) along the Red Sea coast. Indeed, in Grechko's train was Admiral Gorshkov, the commander of the Soviet Navy, who was far more concerned with access to Egyptian facilities than in communicating subtle political messages to Nixon. But Sadat, already in trouble with his military over the existing Soviet bases and the degree of Soviet autonomy exercised, was unwilling to grant further exclusive rights over Ras Banas, although he did allow more Soviet activities there. A single MiG-25 was insufficient inducement. Thus Grechko returned to Moscow, MiG-25 in hand, and succeeded only in further annoying the already irritated Egyptians.

The Nixon-Brezhnev summit in May 1972, was a crucial event, bringing together the frustrations Sadat had with the relationship. The summit demonstrated the subordinate role that patron-client relationships play in superpower affairs. The summit communique gave but a secondary emphasis to the Middle East, dismissing problems in the region as a residual area of confrontation that would soon go away. In the interest of preserving a harmonious beginning to détente, Brezhnev agreed to language that placed the U.S.-Soviet client conflict on the back burner of global political issues. The communique stressed the need for a political settlement, emphasized the importance of mutual restraint, and made no mention of the Palestinian issue. More importantly from Sadat's perspective, the communique called for a "military relaxation" in the region. This, Sadat saw, would freeze Egypt into a position of strategic inferiority vis-à-vis Israel.

To Sadat, this asymmetry in the military balance was Moscow's fault; the realization of his low position in the Soviet political calculus and the demonstrated willingness of Moscow to ignore Egypt's problems in order to make progress in Soviet-U.S. relations were, in Sadat's words, "the straw that broke the camel's back."[55] The promised Soviet explanation of the summit arrived in Cairo a full month after the conclusion of the summit and provided little solace for Sadat. He wrote a seven-point paper in response, demanding that the Soviet Union explain how it would honor its April pledge either to forge a political settlement or to provide Egypt with the offensive weapons Sadat needed to effect a military solution.[56]

Sadat found that the Soviet response to his letter, which arrived in Cairo on July 6, was entirely unsatisfactory. Sadat says that

> more than two and a half pages long, the analysis did not deal with our battle or the weapons needed for it [as was agreed in April, 1972], except in the last five lines which said simply that we were unable to start a battle, that they had experience in this respect, and that they had made an unusual effort to persuade President Nixon that Security Council Resolution 242 should be implemented. . . . There was no mention of ships, the 5 months, the agreement, weapons, or anything like that.[57]

It was now clear to Sadat that the Soviet Union had no intention of helping Egypt with its most pressing national objective: regaining the Sinai from Israel. It was in the Soviet interest, as Heikel argued in a series of *Al Ahram* editorials, to see the "no war, no peace" situation continue indefinitely.[58] Heikel, who basically supported the Soviet position in the Third World, further presented criticisms of Moscow in a mid-June symposium, taking the Soviet Union to task for its reluctance to supply weapons and for permitting the "no war, no peace" situation to continue. Heikel, acting for Sadat, labeled the Soviet Union "criminals" for its behavior in the Middle East.[59]

The callousness with which the Soviet Union treated its Egyptian client in the first half of 1972 was not due to a lack of continued Soviet interest in the relationship; the type III goals of strategic advantage remained as strong as ever. Indeed, Egypt was becoming increasingly important to Gorshkov's Mediterranen strategy, and it was also beginning to figure prominently in the expanding Soviet interest in the Indian Ocean, especially with the possibility of the use of Ras Banas (Berenice) at some point in the future.

Soviet mistreatment of Egypt was, instead, due to three principal factors. First, Moscow believed, as Sadat argues, that "they had Egypt in their pocket."[60] Brezhnev recognized that the "no war, no peace" situation was, in fact, maintaining a high-threat environment for Sadat without risking any military embarrassment. By maintainng a high-threat environment, Moscow felt that it was assuring itself continued access to Egyptian facilities at low risk to its own interests.

Second, Moscow saw the Treaty of Friendship and Cooperation as a formal guarantee of its position in Egypt. The treaty was considered a major ingredient of Soviet regional diplomacy; the skepticism with which such treaties became viewed later was not yet in evidence. The Soviet Union felt that Sadat had no alternative patron to whom he could turn for military and political support in dealing with Israel.

Third, and most important, the Soviet Union began to see that the tactic of détente could yield substantial gains in reducing the strategic nuclear superiority of the U.S. The SALT process was about to produce an anti–ballistic missile treaty and other limiting measures that promised to provide Moscow time to close the strategic gap, assume a position of parity, and obtain recognition as a superpower equal to the United States with a right to have a voice on all international issues.[61] As important as it was for the Soviet Union to extend its naval and political reach into the Mediterranean and beyond, Moscow was unwilling to risk its more critical strategic objectives in an Arab-Israeli war. It must be remembered that patron-client state relationships are but tools in a competitive international environment; if a superpower finds other means with which it can compete with its rivals more effectively, a particular patron-client state relationship may be easily sacrificed.

Taken together, these factors meant that the Soviet Union calculated that the risks to its relationship with Egypt, incurred by an unwillingness to meet Sadat's demands, were small. But even had they been substantially greater, the Soviet Union would have, in all likelihood, taken them anyway, rather than risk the strategic advantages it saw in the evolving détente process. Since both its relationship with Egypt and détente were tools designed to improve Moscow's global position vis-à-vis the United States, it follows that the Soviet Union would prioritize these tools and pursue them accordingly.

However, as in all patron-client state relationships, especially those of type III, the client retains muscle of its own and has the potential to manipulate its patron, notwithstanding the most careful patron calculations.

CRISIS MANIPULATION: THE SOVIET EXPULSION

The events of the first half of 1972, especially the U.S.-Soviet summit, showed Sadat that a major watershed in Egypt's relationship with the Soviet Union was at hand. Egypt had to force the issue with Moscow in a manner that would require, in Sadat's words, "fresh negotiations under the terms of the Treaty."[62] Sadat decided, after the failure of the Saddek mission, to attack the relationship at its most vulnerable point: the Soviet presence in Egypt.

Knowing that Moscow would only react to a dramatic gesture, on July 6, the same day he received the final Soviet responses to his earlier letter, Sadat summoned Soviet ambassador Vinagradov and gave him a five-part instruction to submit to Moscow. The message, dramatic in form and content, said:

> One. I reject this message you have transmitted to me from the Soviet Union. It is unacceptable. I reject, too, this method of dealing with us.
>
> Two. I have decided to dispense with the services of all Soviet military experts [about 15,000] and that they must go back to the Soviet

Union within one week from today. I shall convey this order to the War Minister.

Three. There is Soviet equipment in Egypt—four MiG-25s and a Soviet-manned station for electronic warfare. You shall either sell these to us or take them back to the Soviet Union.

Four. No Soviet-owned equipment should stay in Egypt. Either you sell it to us or withdraw it within a fixed date.

Five. All this should be carried out in a week from now.[63]

Sadat cleverly provided himself yet a little more breathing room, despite the dramatic nature of the announcement. He did not go public with it, nor did he send out the necessary orders right away. Instead, he dispatched Egyptian Foreign Minister Sidqi, a friend of the Soviet Union, to Moscow to see if his announcement had any positive effect. But when Sidqi reported back that no progress had been made, Sadat carried out his threat.

Using the forum of an July 18 address to the Arab Socialist Union, Sadat announced that:

After studying the situation in all its aspects in full appreciation of the tremendous aid the Soviet Union has extended to us, and while fully anxious for the friendship of the Soviet Union, I found it appropriate, as we are on the threshold of a new state of this friendship, to make the following decision:

1. To terminate as from 17 July, the mission of the Soviet military advisers and experts who came here in compliance with our request. Our men in the Armed Forces are to replace them in all the work that they were doing.

2. All the military establishments and equipment which were set up on Egyptian soil in the period that followed the June 1967 aggression will become the exclusive property of the Arab Republic of Egypt and will fall under the adminstration of our Armed Forces.

3. To call for an Egypt-Soviet meeting on a level to be agreed to exchange views with regard to the coming stage. This should be done within the framework of the Cooperation and Friendship Treaty with the Soviet Union.

The first and second resolutions have been put into force as from yesterday.[64]

The Arab Socialist Union (ASU), the Soviet Union, and the entire international community were stunned by this announcement. Moscow never expected that Sadat would take such action, even after his message to Vinagradov and Sidqi's trip to Moscow. But having not responded to Sadat's initial challenge, there was little the Soviet Union could do except comply. There were no Soviet divisions nearby with which Moscow could extend the Brezhnev Doctrine to Egypt, nor was there an effective internal force which the Soviet Union could rely upon to overthrow Sadat.

In the face of this dramatic move, *Tass* announced:

> Now the Soviet military personnel in the Arab Republic of Egypt have completed their functions. With the awareness of this, after the exchange of opinions, the sides deemed it expedient to bring back to the Soviet Union the military personnel that had been sent to Egypt for a limited period. These personnel will shortly return to the USSR.[65]

It is important to note however, that Sadat did not expel the Soviet Navy from its bases at Alexandria, Port Said, and Mersa Metruh, although he did require some modicum of Egyptian control over Soviet activities and the replacement of the Soviet flag by that of Egypt.[66]

In speeches made after July 18, Sadat stressed that the reason for the expulsion of the Soviet advisers was the unwillingness of the Soviet Union to respond to Egypt's weapons needs. There was no doubt, therefore, that a restructuring of the relationship had to take place with the premise that the Soviet Union would adopt a new policy on the transfer of the needed weapons to Egypt.

Fast on the heels of his expulsion of the Soviet advisers, Sadat made public an agreement between Egypt and Libya for a complete union of the two states. Despite the cyclical nature of such announcements, this was of particular concern to the Soviet Union for two reasons. First, the theocratic Qaddafhi was at the time renowned for his anti-Soviet views, based primarily on the incompatibility of communism with Islam and on the charges that the Soviet Union was engaged in expansionism and imperialism in the Middle East. Second, Libya had recently discovered and had begun to exploit vast new oil reserves. With some $2.7 billion in foreign currency exchanges on hand already, and a guaranteed income of at least that much in the future, Libya could make a substantial contribution to the ability of Egypt to procure weapons elsewhere. Europe, unfettered with ideological commitments to Israel, was a possible source, especially given Egypt's overt break with the Soviet Union .

As a result of all these actions, relations between patron and client reached a nadir by mid-August 1972. Sadat had done what he had dared not do in the past; he had tampered with the basic nature of the type III relationships. By reducing Soviet access to Egyptian facilities, Sadat had put at risk the most fundamental Soviet objectives in the relationship and had, therefore, put at risk the relationship itself. In doing so, he had created a crisis of the first magnitude, and only Soviet acquiescence would allow the relationship to continue. This meant that the Soviet Union would have to provide Egypt with the weapons it desired and underwrite Sadat's military plans.

But, in August, Moscow was not yet prepared to make such a decision. The war of words, which began in early August, precluded any rapprochement.

Brezhnev himself responded to a mounting crescendo of rhetoric in a letter to Sadat in which he was far from conciliatory:

> Where is Egypt going? Where is it being driven by forces inside and outside its borders? What is the relationship between us to be in the future? These are the questions which are causing anxiety to your friends and giving encouragement to your enemies. We look forward to receiving an answer to these questions and hope that it will be made in all frankness.[67]

Of all the various correspondence between the Soviet Union and Egypt, this was probaly the most honest. The Soviet leadership was genuinely confused by what it saw in Egypt and perplexed as to the extent of the breach in the relationship. Patrons often have difficulty understanding their clients who operate in different frames of reference.

By the end of August, Sadat grew tired of the bitter exchanges between himself and Brezhnev and between the presses of the respective countries. He decided to lay out, in definitive letter form, his perspective of the Soviet-Egyptian relationship. In this letter, Sadat established a date of October 15 by which the Soviet Union had to "show its good intentions and respond to our demands. [Otherwise] I would be free to take the decisions I deemed fit."[68]

Although Sadat had derived considerable domestic benefit from his ejection of the Soviet advisers, his threat environment remained high. His dramatic actions and subsequent rhetoric had done nothing to alleviate his basic problem with Israel, nor were his efforts at a rapprochement with the West bearing fruit. Sadat was unfortunate in his timing if he sought to effect a new relationship with the United States. Nixon, in the heat of an election campaign, was in no position to begin closer ties with Egypt, a move that would cost him votes with the supporters of Israel in the United States. In addition, Nixon had no desire to ruffle Soviet feathers any more than was necessary at a time when he was attempting to elicit Soviet support for his efforts to extract the United States from Vietnam.

Sadat's efforts to find military support in Western Europe were also running into serious trouble. Foreign Minister Zayyat, in August, arranged for a trip through several European capitals, a trip that he felt would result in a major foreign policy shift for Egypt. But the attack on the Israeli athletes by Palestinian terrorists in Munich in October effectively ended Zayyat's chances. Trips to Germany and Italy were cancelled at the request of the host governments, and Zayyat got a very chilly reception in Britain, which turned frigid when a Palestinian letter bomb killed the Israeli agricultural attache in London.

At the same time this rapprochement effort with the West fizzled, Sadat came to face the realities of his military capability to reduce his threat environment with no help from the Soviet Union. Immediately following the expulson of the Soviet advisers, Sadat had directed Saddek to prepare plans to attack the Sinai in November. He felt that the Egyptian armed forces, although not ready to recapture the entire Sinai, would be able to win a limited engagement against Israel, and parlay that into a major political success. Saddek's initial response to Sadat was that such an attack was, indeed, feasible, and that it could be ready by the time frame Sadat had specified. But, in late September, Saddek, with the support of the major Egyptian commanders, told Sadat that such an attack was impossible without additional weapons.

All this meant that Sadat once again had to turn back to the Soviet Union. Indeed, it is improbable that Sadat ever contemplated any other outcome to his manipulated crisis. His expulsion of the Soviet advisers was a tactical move designed to force the Soviet Union to provide the offensive weapons he needed and to placate anti-Soviet sentiment within Egypt. If Sadat could have effected a political rapprochement with the West, that would have been an added, if unintended, benefit. Indeed, Sadat had little faith in the ability of the United States to bring about a political solution to the Sinai that was acceptable to Egypt. He remained confident that only military action would be productive, at least until Egypt's credibility as a military power was reestablished.

The Soviet Union, for its part, had been stunned by Sadat's expulsion of its advisers and by the subsequent rhetoric that poured forth from Cario. Ever suspicious, Moscow was convinced that there had been collusion between Sadat and the United States in the sudden chilling in the Soviet-Egyptian relationship.[69] The Soviet Union had no desire to see the relationship end; it had invested an awesome amount in it and needed the strategic benefits derived from it.

The Sadat letter of August 29 stimulated a major review of the relationship within the Soviet Politburo. As with all crises in patron-client state relationships, the expulsion of the Soviet advisers had clearly identified what the stakes were and the nature of the response that was required if the relationship were to survive.

Further weighing in the Soviet calculus was the forthcoming renewal of the naval access agreement. This agreement, signed in Janauary 1968, had a five-year period, and the renewal was due in January 1973, a few monthes hence. Sadat had carefully excluded Soviet naval presence in Egypt from his expulsion order, and now the prospects of losing naval access loomed heavily on the horizon.

These factors taken together meant that, by October 1972, patron and client were prepared to patch up the relationship. With this as a backdrop, Sadat dispatched Prime Minister Sidqi to Moscow almost exactly three months after the expulsion order to ascertain how his manipulation of the crisis had worked. The Sidqi visit provided the catalyst to get the relationship going again. The visit began on a cautious note and did not result, by itself, in a resumption of the flow of arms. But the visit was hailed in both capitals as a major success, and established the momentum for more constructive relations in the immediate future.

Sadat moved quickly to prove his good intentions. Shortly after Sidqi returned from Moscow, Sadat fired Mohammed Saddek. to the delight of the Soviet Union. The Soviet press hailed this decision as a major gesture on the part of Sadat and urged that closer ties be the objective in the immediate future.[70]

The combination of the Sidqi visit and the Saddek firing was sufficient to resume the flow of arms. The first deliveries, brought by a Soviet military mission in November 1972, contained SA-6 missiles, the mobile surface-to-air missile so needed by Sadat in order to effect even a limited offensive operation against the Israelis in the Sinai. The SA-6 had been in Egypt before but always under the control of the Soviet forces.

Sadat solidified the rapprochement shortly after the arrival of the SA-6 batteries by announcing a five-year extension of the naval access agreement. In doing this he responded to, and reaffirmed, the most fundamental Soviet goal in the relationship. This decision was probably the final gesture required to convince Moscow to resume the relationship on Sadat's terms. Indeed, by the end of the year, the flow of arms into Cario became a flood. Sadat, in an unaccustomed position of trying to assimilate a substantial influx of weapons, speculated that "they [the Soviets] want to push me into battle."[71] Sadat's statement, while overdrawn, was essentially correct. The firing of Saddek and the decision to renew the naval access agreement convinced the Soviet Union of the correctness of its October Politburo decision that, even if it risked a resumption of the war, the relationship was worth the cost.

The manipulated crisis of 1972, therefore, resulted in an almost complete Egyptian victory. Sadat had fully recognized that, without such a crisis, Soviet support would have continued at a wholly unsatisfactory level. By precipitating a crisis in his relationship with the Soviet Union, and by choosing a mechanism that attacked Moscow's most basic goals, Sadat forced the Soviet Union to increase its level of support for Egypt in a manner that met Sadat's demands.

Sadat's crisis manipulation, although successful, was not entirely the model of Machiavellian brilliance as it is sometimes portrayed.[72] Both Egypt and the Soviet Union miscalculated in the 1972 crisis, a characteristic of all crises in patron-client state relationships. Moscow felt that, just as in 1971, Sadat was bluffing in this threat to expel the Soviet advisers. Even after Sadat told Vinagradov on July 6 that he was prepared to make such a move, the Soviet Union did not believe him and was, therefore, unresponsive to the Sidqi visit eight days later. After the expulsion order, the Soviet Union was thrown into confusion over what was happening in Egypt, as evidenced by Brezhnev's mid-August letter to Sadat.

Sadat, for his part, overestimated the readiness of his armed forces to resume a war against Israel. Although his primary goal in the crisis was to force additional support from the Soviet Union, he was reasonably confident in July that, even if his gambit failed and the Soviet-Egyptian relationship died, he would still be in a position to engage Israel in a limited war with some chance of success.

When both sides recognized the extent of their miscalculations they hastened to patch up the relationship but on Sadat's terms. Indeed, the rapidity with which the Politburo decided on a major change in policy, that not only affected its relationship with Egypt but that had certain implications for its relationship with the United States, speaks of the success of Sadat's crisis manipulation.

The net result of Sadat's manipulated crisis of 1972 was a tacit acceptance by Moscow of Sadat's plans to embark on a limited war against Israel in 1973. This reversed the long-standing Soviet opposition to a resumption of hostilities. The lesson to be learned from this experience is that crisis manipulation is more likely to succeed if the manipulator attacks his partner's most basic objectives in the relationship. The Soviet Union did this in 1967; Sadat did it in 1972.

PRELUDE TO WAR

Although relations between the Soviet Union and Egypt continued to improve in early 1973, and there was tacit agreement between the two states that Egypt would resume the war sometime during the year, Sadat's problems were by no means over. His pronouncement, in January 1973, that Egypt would resume the war soon was greeted with considerable skepticism by a population that had lived the "year of decision." Moreover, anti-Soviet feelings remained high, as the Egyptian people did not know that this time the Soviet Union really was responding to Sadat's requirements. There was continued restiveness in the army, and the students once again took to the streets.

Nonetheless, Sadat began vigorous preparations to fight Israel in the Sinai. He continued to court Saudi Arabia's King Feisel in an effort to get him to use his oil to influence the West. In fact, Saudi money paid for many of the weapons provided to Egypt during 1973, helping to relieve Soviet cash flow problems.[73] Sadat also solidifed relations with Kuwait, Lebanon, Morocco, Algeria, and Tunisia.[74]

In Egypt, Sadat faced mounting domestic pressure, this time emanating from the Left. With rumors of still another coup in the making, Sadat fired Prime Minister Sidqi and purged the Arab Socialist Union of the "adventurist left."[75] He assumed the position of prime minister himself and proceeded to abolish what was left of Arab socialism in Egypt. All of this domestic turmoil was of great concern to Sadat; his position increasingly depended upon his ability to resume the war with Israel.

Despite the turmoil, the shipment of Soviet arms continued at a brisk pace throughout the early months of 1973. Sadat knew well that there was a considerable time lag between the arrival of weapons and their effective use on the battlefield. Time was required for individual soldiers to be trained on new weapons not only so that they would know the technical operations and maintenance procedures, but also so that they could gain confidence in their use and be able to survive in battle. Therefore, it was of great importance to Egyptian military operations that Soviet weapons arrive early and in quantity.

The weapons provided to Egypt answered all of Sadat's needs except for one. The Soviet Union provided Egypt with quantities of SA-6 and SA-7 mobile air defense weapons; AT-3 Sagger antitank guided missiles to defeat Israeli armor; 90 T-62 tanks to enhance and upgrade Egypt's armor forces; SS-C-26 Samlet naval missiles to neutralize the limited Israeli Navy; a Su-20 Fitter ground attack aircraft with which to attack Israeli troop concentrations massing in the Sinai. The one weapon not provided was Sadat's much desired deterrent weapon; no MiG-23 or MiG-25 aircraft were delivered to Egypt, nor was the medium-range surface-to-surface SCUD missile provided.[76] These Sadat needed in order to provide a credible deterrent to Israeli attacks on Egyptian population centers. In this regard, it was less important to Sadat that Egyptians man these weapons, for he never intended to use them.

In order to move Moscow on this last of the needed weapons, Sadat initiated yet another mini-crisis. As the second Nixon-Brezhnev summit approached in May 1973, Sadat told Moscow that he intended to begin the war immediately, knowing full well that, in the absence of a deterrent weapon, this was impossible.[77] This threatened to derail the summit and throw the path of détente askew. Soviet concern was further reinforced by the venue of the summit; it was to be held in Washington. The dangers of an uncontrolled client would be intensified if a war occurred while Brezhnev was in the U.S. capital.

Moscow was, therefore, probably willing to strike a deal with Sadat. In return for calling off an attack he never intended to carry out, Sadat received a promise of SCUD missiles.

Sadat's mini-crisis of May 1973 served a tactical purpose quite apart from the issue of the deterrent weapon. As part of a highly effective general deception plan, he allowed the news of his pending May attack to leak to Israel. Sadat's intention was to force Israel to mobilize.

> As part of my strategic deception plan, I launched a mass media campaign [of which the *Newsweek* article was a part] which led the Israelis to believe that the war was imminent. On the days when war seemed likely to break out, there was a full Israeli mobilization while we enjoyed perfect military calm. I did the same thing in August—and the Israeli reaction was the same. After the October War, Moshe Dayan was asked why he hadn't mobilized in October. He said that Sadat "made me do it twice, at a cost of ten million dollars each time. So, when it was the third time around, I thought he wasn't serious."[78]

In truth, Sadat was not ready to go to war in May. The Soviet Union had only provided a 90-day supply of spare parts and ammunition, a thin margin on which to base even a limited conflict. Moreover, the deterrent weapon had not yet been deployed, and Egyptian forces still required additional training on the new weapons systems.

As the Nixon-Brezhnev summit materialized in 1973, Sadat again faced the problem of being placed on a back burner while grander issues of superpower relations were discussed. Prior to the summit, Sadat sent Ismail to Moscow to urge Brezhnev not to sacrifice Egyptian interests in the name of détente. His interventions with Brezhnev came to naught, however; the summit communique contained no substantive references to the Middle East, nor was there any progress in resolving the Israeli occupation of the Sinai.

But in 1973 Sadat was far less concerned with this outcome than he had been the year before. The Soviet Union was providing the weapons he required, and it was only now a matter of operational timing before Egypt could resume the war against Israel.

Indeed, by mid-summer, the Soviet Union had resigned itself to a resumption of the war. In April Kosygin had said in a Stockholm interview that "Egypt has the right to possess a powerful army now in order to liberate its own lands."[79] At the July 30–31 meeting of the Warsaw Pact political leaders, this message was reaffirmed. The summit communique asserted that the Middle East problem could be solved only by "full withdrawal of Israeli forces from the occupied Arab territories." Gone were the usual caveats about a peaceful solution and any references to U.N. Resolution 242.[80]

Because the nature of the Soviet-Egyptian relationship was so enmeshed in the U.S.-Soviet relationship, the willingness of Moscow to allow Egypt to proceed along a military road was, to some extent, reflective of the state of détente in 1973.

By mid-1973, it was apparent to the Soviet Union that détente was not going to yield all the immediate benefits that had been anticipated the year before. The granting of "most favored nation" status, which would have permitted a considerable trade advantage for the Soviet Union, was not yet a reality, and there were indications that it would experience substantial complications. Efforts to tie this agreement to a loosening of Soviet emigration rules for Soviet Jews desiring to move to Israel were developing in U.S. circles.

In addition, the strategic arms limitation talks gave evidence of bogging down and showed themselves to be, at a minimum, a long and arduous process.

Moreover, the mounting domestic turmoil surrounding U.S. President Nixon caused concern in Moscow that the détente for which the Soviet leadership felt Nixon was personally responsible, would evaporate if he were unable to continue in office.

Equally importantly, Sadat had uncovered a potentially effective weapon that, if properly used, could result in major strategic gains for the Soviet vis-à-vis the United States. In July Sadat had obtained a warning from King Feisel to the United States that Saudi Arabia would not increase its production of oil unless the U.S. brought pressure to bear on Israel to make significant concessions in the Sinai. This demonstrated to both Egypt and the Soviet Union the potential that oil held as a weapon against the United States and its more vulnerable NATO allies.[81]

Thus, given the pressures Sadat had applied during the expulsion crisis and the lessening of the glow of détente, the Soviet Union was prepared to allow Egypt to embark on combat operations against Israel.

On September 12, the deterrent weapon finally arrived in Egypt; the Soviet Union moved two brigades of SCUD C surface-to-surface missiles to Egypt. As agreed by Sadat, the SCUD units remained under strict Soviet control. Since this was intended to be only a deterrent weapon, such as arrangement was acceptable to Sadat.[82]

Committing the SCUD brigade to Egypt was the final act Sadat needed before proceeding with the war against Israel. Moscow knew full well the import of the SCUD; without it, Sadat might have deferred his decision yet a while longer. The Soviet deployment thus represented the complete victory of Sadat over his Soviet patron, a victory rooted in the successful crisis manipulation of the year before.

From that point on, the preparations for war continued at an accelerated pace. By mid-September, two Egyptian armies (the Third and the Second) were deployed along the Suez Canal in prepared regimental camps.[83] SA-6 sites, constructed during the spring and summer months, became operational. Soviet bridging equipment, including Soviet-made heavy bridges designed to carry tanks, was stockpiled in visible locations along the entire Canal front. At the same time, a multi-division Syrian force moved into the Golan Heights into positions opposite Israeli units.

On October 3, Sadat summoned Vinagradov and informed him that "I and Syria have decided to start military operations against Israel so as to break the present deadlock. What will the Soviet attitude be?"[84] Sadat and Syrian president Hafez Assad had agreed that the actual date for beginnings of hostilities would be given to the Soviet Union the following day in Damascus. The announcement to the Soviet Union did not come as any great surprise; since the delivery of the SCUD units to Egypt, Moscow knew that it was only a matter of time before Egypt would cross the Canal and attack Israel positions in the Sinai. Nonetheless, Sadat felt constrained to inform his Soviet patron and, at least ostensibly, give Moscow one final chance to weigh in against the attack.

The following day, Sadat received his answer. Vinagradov communicated an urgent request from Moscow that the Soviet Union be allowed to evacuate dependent personnel from Egypt before the outbreak of hostilities. Sadat, of course, granted permission, although he felt that this was the final Soviet display of contempt for the Egyptian military.

The evacuation of Soviet noncombatants from Egypt began in earnest on October 5, as U.S. intelligence confirmed that civilians and nonessential personnel were being shipped out of Cairo on Aeroflot aircraft. Some of these aircraft were diverted enroute to other locations and ordered to fly to the Middle East.

The following day, the Soviet Union launched a photo-reconnaissance satellite that tracked across the Middle East. Indeed, between October 3 and 17, the Soviet Union launched five such satellites. Moscow intended to watch the war very carefully.

On the morning of October 6, Egyptian and Syrian divisions attacked on two fronts in a limited war that was to have a profound impact on the future of the Middle East and the relations between the superpowers themselves. The management of the Egyptian-Soviet patron-client state relationship during this major international crisis reveals important insight into the nature of such relationships in general. Because of this, it is necessary to examine the behavior of both patron and client in some operational detail. This will be developed further in the next chapter.

During the precrisis period of 1973, the relationship was clearly dominated by Egypt, although the relationship remained type III in nature. The Soviet Union acquiesced to Egyptian demands, despite the risks they entailed for broader Soviet interests. Soviet policy during this period became one of damage limitation; Moscow wanted to ensure that, if Sadat went ahead, the conflict would be limited in scope and would not result in a U.S.-Soviet confrontation, nor in another humiliation for the Arab world.

CONCLUSION

The relationship between the Soviet Union and Egypt from 1967 to 1973 shows the dynamic nature of the flow of influence in patron-client state relationships and the various mechanisms both states use to manage the relationship. It also illustrates the fundamental objectives each side has in such relationships and how these objectives shape events.

As long as the Soviet Union maintained goals of international solidarity in the relationship, Moscow tended to be the dominant partner. This was particularly true in the period immediately following the 1956 Suez Crisis, when Nasser faced a period of high threat. But, as soon as the Soviet calculus and objectives changed as a result of events in the mid-1960s, and the relationship moved to type III, the dynamics of the flow of influence became more complex, with dominance shifting back and forth depending upon the degree of Egypt's threat environment and the political maturation of Anwar Sadat. In periods of high threat, such as during the war of attrition, Soviet dominance reasserted itself; in periods of reduced threat, Egypt often gained the upper hand.

Both sides used, to differing degrees of success, crisis manipulation to achieve certain objectives in the relationship. Periodically, the Soviet Union, through its actions, demonstrated the subordination of its relationship with Egypt to more fundamental relations with the United States. These efforts were most instrumental in recognizing that the Soviet-Egyptian relationship was important to the Soviet Union inasmuch as it served broader Soviet objectives vis-à-vis the United States.

The lessons that can be derived from a systematic understanding of the Soviet-Egyptian patron-client state relationship are of substantial importance not only for heuristic purposes but also for their policy implications. The Soviet-Egyptian relationship is reflective of the underlying nature of other patron-client state relationships. As the international system faces periods of expanding competition in which patron-client state relationships will serve as surrogates for direct confrontation, understanding these relationships will be of crucial importance to the effective management of international affairs.

NOTES

1. Cited in Jon D. Glassman, *Arms for the Arabs* (Baltimore: Johns Hopkins University Press, 1975), p. 39.

2. Glassman, *Arms for the Arabs*, p. 39. This sort of "have you stopped beating your wife" accusation could have had no effect other than to stimulate Nasser to show it was not true.

3. Anwar as-Sadat, *In Search of Identity*, (New York: Harper & Row, 1977), p. 163. Sadat tries manfully to find some good in the Yemen war. He says that "it curbed Saud's aggressiveness." P.J. Vatikiotis, "The Soviet Union and Egypt: The Nasser Years." In Ivo Lederer and Wayne S. Vucinich, *The Soviet Union and the Middle East* (Stanford: The Hoover Institution Press, 1974) p. 129.

4. Sadat, *In Search of Identity*, p. 172; Mohammed Heikel, *The Sphinx and the Commissar*, (New York: Harper & Row, 1978), p. 174. There is no evidence to support the Soviet claim. Indeed, the Israeli plan, as it unfolded, placed the bulk of Israel's limited manpower along the southern front, to be redeployed to the Golan Heights after the Sinai had been taken.

5. Cited in Aryeh Yodfat, *Arab Politics in the Soviet Mirror*, (Jerusalem: Israel Universities Press, 1973), p. 263.

6. Sadat, *In Search of Identity*, p. 172.

7. Sadat, *In Search of Identity*, p. 172. The issue of the Strait of Tiran had been a particularly troubling one, being the source of what Sadat calls "Arab oneupsmanship." Arab states accused Nasser of aiding Israel by allowing the Strait to remain open.

8. Nasser, by contrast, was sure that the Arab states would win. Sadat, *In Search of Identity*, p. 172.

9. Cited in Glassman, *Arms for the Arabs*, p. 41.

10. Ibid.

11. Cited in Yodfat, *Arab Politics in the Soviet Mirror*, p. 264. This was evidently caused by a last statement given to Badran by Grechko, as the former left Moscow in late May. Grechko had, according to Heikel, simply given Badran "one for the road," to boost his morale. Heikel, *The Sphinx and the Commissar*, pp. 179-80.

12. Heikel, *The Sphinx and the Commissar*, p. 178.

13. Badran was promised that, should the U.S. Sixth Fleet intervene, "we will deal them a lethal blow." Sadat, *In Search of Identity*, p. 173.

14. Sadat, *In Search of Identity*, p. 175; Alvin Z. Rubinstein, *Red Star on the Nile*, (Princeton: Princeton University Press, 1977), p. 11. Although no one of substance accepted this claim, Heikel reports that it was widely believed in the streets of Cairo; the Egyptian people simply could not understand how their vaunted armed forces could have been beaten without superpower intervention. Heikel, *The Sphinx and the Commissar*, p. 181.

15. Sadat, *In Search of Identity*, p. 186.

16. *U.S. News and World Report* , January 19, 1970, p. 37.

17. *Christian Science Monitor*, March 18, 1970.

18. Cited in Glassman, *Arms for the Arabs*, p. 80.

19. Edward R. Sheehan, "Why Sadat Packed Off the Russians," in *The New York Times Magazine*, August 6, 1972, p. 41. Sadat was appointed periodically to maintain informal ties with the United States, where he was well liked and respected as a moderate Egyptian in a government of revolutionaries.

20. Included in the Kosygin party were Marshal Zakhorov (the former gray eminence of Egyptian rearmament) and V.M. Vinagradov (soon to be Moscow's ambassador in Cairo).

21. *Economist* (London), September 19, 1970.

22. Glassman, *Arms for the Arabs*, p. 83.

23. Mohammed Heikel, *The Road to Ramadan*, (NY: Quadrangle, 1975) p. 113.

24. Sadat, *In Search of Identity*, p. 210.

25. Sadat, *In Search of Identity*, p. 215. Henry Kissinger shared this view. Henry Kissinger, *The White House Years* (Boston: Little, Brown and Company, 1979) pp. 1276-1277.

26. Heikel, *The Road to Ramadan*, p. 115.

27. Dan Kurzman in *Saturday Review*, September 5, 1970, p. 30.

28. Heikel, *The Sphinx and the Commissar*, p. 29. Heikel is generally more sympathetic towards Sabry than is Sadat. But even he acknowledges that the Soviet Union saw Sabry as its man in Cairo.

29. Heikel, *The Road to Ramadan*, pp. 166–67.

30. Accounts differ regarding which aircraft Sadat sought. At various places, the "deterrent" aircraft is referred to as the MiG-25, the MiG-23 Flogger, and the SU-17 Fitter C. The confusion probably stems from the fact that the Egyptians really used the Soviet designation for the aircraft, which at that time was the X-500.

31. Sadat, *In Search of Identity*, p. 221.

32. Heikel, *The Road to Ramadan*, p. 119. Egypt received some SAM batteries in April 1971 and some of the requested ammunition. Sadat, *In Search of Identity*, p. 221.

33. Anwar Sadat, "The Ice Thaws Between Moscow and Cairo," in *October* (in Arabic), December 12, 1976.

34. Arnold Hottinger, "Soviet Influence in the Middle East," in *Problems of Communism*, March-April 1975, p. 72.

35. Robert O. Freedman, *Soviet Policy Toward the Middle East Since 1970*, (NY: Praeger, 1976) p. 51.

36. Sadat was sincere in his warning of the pending ouster of Sabry to Vinagradov in April; he did not want this personal issue to interfere with his relations with Moscow.

37. Sadat, *In Search of Identity*, p. 225.

38. Article 8 of the treaty, in which this language appears, is unique among the various Soviet treaties throughout the Third World. Only the treaty with Somalia, signed in July 1974, contains any similar language, and there it is clearly related to defensive use of the Soviet-supplied equipment.

39. Glassman, *Arms for the Arabs*, p. 89.

40. Rubinstein, *Red Star on the Nile*, p. 156.

41. Anwar Sadat, "From the Memoirs of President as-Sadat," in *October* (in Arabic), December 26, 1976.

42. Sadat, *In Search of Identity*, p. 227.

43. John K. Cooley, "The Shifting Sands of Arab Communism," in *Problems of Communism*, March-April 1975, p. 24.

44. Freedman, *Soviet Policy Toward the Middle East Since 1970*, p. 66. The immigration issue demonstrated repeatedly the subordination of the Soviet-Egyptian relationship to U.S.-Soviet issues, at least in Moscow's calculus.

45. Heikel, *The Road to Ramadan*, p. 169.

46. Sadat, *In Search of Identity*, p. 228.

47. This was largely disingenuous. In fact, the joint communique issued after the meeting was almost mute on the issue of arms supplies.

48. *Christian Science Monitor*, February 18, 1972.

49. These were the students arrested during the wave of anti-Soviet demonstrations that swept Cairo in January 1972. To be sure, other reasons, such as shortages of staples, led to the demonstrations, but their primary emphasis was against Soviet dominance in Egyptian affairs.

50. Sadat, *In Search of Identity*, p. 229.

51. Ibid.

52. *New York Times*, April 29, 1972.

53. Once again, there is confusion as to the aircraft brought with Grechko; Sadat says that it was the Su-17 Fitter. Most others are convinced that it was the MiG-25 Foxbat (Sadat, *In Search of Identity*, p. 228. Heikel, *The Road to Ramadan*, p. 163. Rubinstein, *Red Star on the Nile*, p. 182).

54. Sadat, *In Search of Identity*, p. 228.

55. Sadat, in a speech on the anniversary of Nasser's death, carried by Cairo Domestic Service, September 28, 1975.

56. Glassman, *Arms for the Arabs*, p. 94.

57. Sadat, "From the Memoirs of President as-Sadat," in *October* (In Arabic), December 26, 1976.

58. Heikel, *The Road to Ramadan*, p. 164.

59. Sheehan, "Why Sadat Packed Off the Russians," p. 51.

60. Sadat, *In Search of Identity*, p. 231.

61. William E. Odom, "Whither the Soviet Union," in *The Washington Quarterly* (Spring 1981), pp. 30–31.

62. Heikel, *The Road to Ramadan*, p. 170. Here again, the differences in interpretation of the treaty were evident. Sadat still believed that the treaty was his carte blanche to redress the humiliation of the Israeli occupation of the Sinai. Moscow, obviously, saw it differently.

63. Sadat, *In Search of Identity*, p. 230.

64. Rubinstein, *Red Star on the Nile*, pp. 188–89.

65. *TASS International Service*, July 18, 1972.

66. Rubinstien, *Red Star on the Nile*, pp. 190-191.

67. Heikel, *The Road to Ramadan*, p. 176.

68. Sadat, *In Search of Identity*, p. 234.

69. This theme was picked up in the Soviet press througout the period.

70. George Lenczowski, "Egypt and the Soviet Exodus," in *Current History*, January 1973, pp. 15–16.

71. Heikel, *The Sphinx and the Commissar*, p. 254.

72. Particularly by Sadat. Sadat, *In Search of Identity*, pp. 230–31. He also says that he threw the Soviets out in order to get them out of the line of fire when the war resumed.

73. Rubinstein, *Red Star on the Nile*, p. 242.

74. Sadat, *In Search of Identity*, p. 239.

75. Cooley, "The Shifting Sands of Arab Communism," p. 25.

76. Glassman, *Arms for the Arabs*, pp. 109–10.

77. Sadat hints at this subterfuge in his September 22, 1974 interview, carried by the Cairo Middle East News Agency.

78. Sadat, *In Search of Identity*, pp. 241–242.

79. Cited in Kohler, *The Soviet Union and the October, 1973, Middle East War*, p. 36.

80. Glassman, *Arms for the Arabs*, p. 101.

81. Freedman, *Soviet Policy Toward the Middle East Since 1970*, p. 112.

82. *Aviation Week and Space Technology*, November 5, 1973, p. 12.

83. The Egyptian Army was organized along Soviet lines; hence an "army" was, in fact, normally composed of but three to four divisions, roughly equivalent to a U.S. corps.

84. Sadat, *In Search of Identity*, p. 246.

CHAPTER 6

CRISIS MANAGEMENT: THE OCTOBER WAR

INTRODUCTION

Crisis management in patron-client state relationships is a substantially more complex problem than is crisis management in interpatron relations. In "level two" crises, that is, those involving patrons only, the antagonists have developed certain implicit and explicit guidelines for resolving the conflict short of nuclear war.

The same cannot be said about patron-client crises. Not only are there few guidelines for managing such crises, but the client is unlikely to understand or share the patron's concern for crisis control. The client, acting in a seemingly capricious and irresponsible way, shapes the crisis often without regard to the desires of the patron.

The management of patron-client crises is thus far more complex than the management of crises involving patrons alone. This is due to the addition into the crisis-management equation of third and fourth parties, parties that do not share common world visions with their patrons. The patrons can no longer control the scope and intensity of the crisis, and the risk of loss of control by the patrons becomes much greater. Each patron must at once try to manage the client, the client's enemies, and the other patron. Often, this involves trying to implement measures that are, in themselves, contradictory. A patron decision to supply arms to its client on an expeditious basis, for example, may please the client but enrage the other patron. The crisis management equation thus goes from the simple linear case of a bilateral patron crisis to the quadratic of a patron-client crisis.

Patron-client crises generally proceed on two levels: that of the clients (level one) and that of the patrons (level two). As a general rule, the patrons will seek to manage the crisis at the lowest level possible. This does not mean necessarily that the patrons will not become actively involved in the client crisis, but it does mean that the patrons will try to manage the client crisis so that it does not escalate to the patron level. But, because of the usual difficulties the patrons experience in managing unruly clients in a crisis environment, this effort often proves fruitless.

There are, of course, patron-client crises that do not involve other patrons. These are, in themselves, difficult to manage but do not carry with them the degree of threat to the international system that crises involving clients of more than one patron do.

Given these problems in managing patron-client relationship crises, it is no surprise that such crises often get out of control and find the patron struggling in an ad hoc, improvised fashion for a resolution of the crisis, sometimes with potentially disastrous results.

PATRON AND CLIENT GOALS IN THE CONFLICT

The October 1973 war between Israel and its Arab neighbors is an important example of the implications of patron-client crises both in terms of the relationship itself and for the international system as a whole. In the span of three weeks, fundamental changes were wrought in the region, in East-West relations, and in the international economic order. These changes were largely generated by the loss of patron control in the client crisis, by the escalation of the conflict to superpower levels, and by the unexpected developments on the battlefield.[1]

There is some justification to the claim that the October War was not really a crisis. Indeed, as we saw in the previous chapter, both Sadat and the Soviet Union knew full well that the war was coming, and therefore, the event came as no surprise.[2] But, although the event itself was not a surprise, the way in which the war developed and the demands it placed on both patron and client introduced a strong element of surprise. All patrons and clients were unprepared for the intensity of the conflict and for the tactical developments on the battlefield, and all the peculiarities of crisis decision making became operative. Therefore, it is a justifiable departure from the strict definition to call the October War a crisis.

Moreover, it was perhaps the first true crisis in the Soviet-Egyptian relationship. All others, including the 1967 war, were to some extent manipulated crises brought on by either patron or client and so were designed to have their primary effect on the relationship itself. This was not the case in the October War. Sadat did not begin the war with an objective of influencing Soviet support for Egypt as Nasser had done in the war of attrition; such support had been more or less successfully obtained by the manipulated crisis of July 1972. Sadat was, instead, seeking to meet his fundamental national objective of Egyptian security by reducing his threat environment. Soviet support was, and had always been, a means to this end.

The following discussion of the war is not intended to be a definitive history; perhaps no other war in the past three decades has received such exhaustive historical review, and another such review is not necessary now. But, in order to understand the crisis-management efforts by both patron and client, it is necessary to outline the operational details of the conflict that formed the underpinnings of the management of the crisis and of the escalation to superpower confrontation. No analysis of crisis management in the war can be adequate without inclusion of the highlights of the conflict itself.

Soviet Objectives in the Conflict

When the Soviet Union made its decision, in October 1972, to resume its relationship with Egypt, it did so in the full knowledge that it probably would have to pay Sadat's price of support for his effort to renew the war against Israel. Within this general understanding of the implications of its actions, Moscow sought several important objectives that, in the Soviet view, were attainable.

First, the Soviet Union sought to limit the scope and duration of the conflict. Moscow recognized that the longer the conflict lasted and the more wide-ranging it became, the greater the chances of unpredictable outcomes were. The Soviet Union historically has avoided uncertainty; a limited war was one of the important outcomes to reduce Moscow's uncertainty. The Soviet Union retained its utter contempt for the capabilities of the Egyptian military and was confident that its client could not win even a limited engagement with Israel. This drove Moscow to seek a rapid termination of hositilities as soon as the conflict began. It had no desire to see its client experience yet another humiliation at the hands of the U.S. client, Israel.

Second, the Soviet Union wanted to reestablish the "no war, no peace" situation that had brought it so much success in the relationship. In this regard, the Soviet Union sought an outcome in which there was no decisive victor, either Israel or Egypt. A clearcut defeat of Moscow's Arab clients, and in particular Egypt, would be seen as another demonstration of the superiority of U.S. weapons over Soviet arms and would call into question the efficacy of subordination to Moscow in the eyes of other Third World clients. Moveover, a decisive defeat of the Arab armies would almost assuredly result in anguished cries for direct Soviet intervention, something that in the cold light of precrisis reality, Moscow wanted to avoid.

On the other hand, a decisive defeat of Israel, as unlikely as such an outcome may have been in Soviet eyes, would almost certainly bring about an intervention by the United States. This, in turn, would quickly escalate into a superpower military confrontation, for the Soviet Union could scarcely avoid committing its own forces in response.

Third, the Soviet Union wanted to demonstrate its reliability as a patron. The events of the preceding year had called into question that reliability, and Moscow needed to show that, in a crisis, it would support its client in whatever manner was necessary. The Soviet Union probably hoped that, as in 1967 when the war too short to require any real support during the crisis, the 1973 war would be of sufficiently limited scope that it would allow a demonstration of support to take place after the heat of battle had passed.

Fourth, the Soviet Union wanted to use the pending 1973 conflict to solidify its political and military position in the region. This involved not only the objectives mentioned above, but also the flexibility to exploit opportunities as they might arise in the course of the conflict. Sadat's inroads into the conservative Arab oil bloc were not lost on Moscow, nor was the potentially massive damage oil as a weapon could wreak on the economies of the West. The Soviet Union wanted to be in a position to exploit any favorable, if unexpected, development that might occur, while limiting the risks of superpower confrontation.

Fifth, the Soviet Union wanted to underscore Egypt's dependence on its patron. The 1972 expulsion of the Soviet advisers was a humiliating experience rooted in what Moscow saw as erroneous calculations by Sadat regarding Egypt's need for its relationship with the Soviet Union. In a conflict, even one of limited scope, Egypt's threat environment would rise enormously, and only the Soviet Union would be available to fill the gap. Heikel's report of Grechko's arguments to the Politburo in October 1972, illustrates this point.

> Let the Arabs have sufficient arms to enable them to risk a battle. Should this happen, and should the Arabs win, their victory would have been achieved thanks to Soviet arms. Should they be defeated . . . it is still to the Soviet Union that they will have to look for rescue in the aftermath of the battle.[3]

Grechko remembered well that it was only after the 1967 debacle that the Soviet Union achieved its basing rights in Egypt and that these were expanded only after the Soviet intervention in the war of attrition. Like a step function, expansion of access to Egyptian facilities was closely linked to Soviet reponsiveness to Egyptian crises.

Finally, and most importantly, the Soviet Union sought to avoid a military confrontation with the United States. Although the Soviet Union had greatly expanded the size of its naval presence in the Mediterranean, thanks in part to its Egyptian bases, Moscow recognized that it was still in no position to engage the United States in the region. Moreover, at a time of residual nuclear inferiority, the Soviet Union had even less desire to engage the United States in a general war. This overriding precrisis objective set limits on the pursuit of other opportunistic goals.

All of these objectives were felt to be obtainable by the Soviet Union. But Moscow, well versed in crisis management, knew full well the vagaries and high uncertainties that are the substance of international crises, particularly those involving client states. Coupled with this inherent crisis uncertainty was the degree of control that the Soviet Union knew it was surrendering to Sadat, control that added to the complexities of crisis management.

Soviet Precrisis Management

With these objectives, and the uncertainties inherent in a client crisis in mind, the Soviet Union took a number of steps before the war began to manage the crisis to the extent possible. These measures are sometimes pointed to as evidence of Soviet collusion in the October attack; in fact, they were probably precautionary measures taken to enhance Soviet control over events once the war, a war that Moscow did not want but felt constrained to support, began.

First, as mentioned in Chapter 4, the Soviet Union evacuated its noncombatants on October 4–5. This was done in order to reduce the leverage either Egypt or Israel could exert on Moscow through the exploitation of potential Soviet hostages.[4]

Second, the Soviet Union increased its satellite coverage of the region in order to provide the most accurate battlefield assessments possible. The Soviet leadership knew that its ability to influence events once the crisis began would depend upon intimate knowledge of the state of the war on the ground. It is doubtful that the Soviet Union anticipated using the information gathered by its satellites to contribute to Egypt's tactical intelligence. Soviet satellites were not designed for that purpose, nor did Moscow want to play that active a role in the war.[5]

Third, the Soviet Union made preparations to ship equipment to Egypt and Syria. The shipment of massive amounts of equipment to any conflict is no mean feat; it involves an extraordinary amount of packaging, preparations, and movement, especially when the equipment involves major end items such as tanks or howitzers. The Soviet Union recognized that a major resupply effort, to be effective, would have to be organized early. It is doubtful that the Soviet Union made a commitment, even internally, to embark on a resupply effort. Soviet preparations were more likely designed to provide Moscow with yet another option to be available, should the circumstances dictate that a resupply effort was in the Soviet interest.

Fourth, the Soviet Union directed that certain arms-carrying merchant ships enroute to Egypt stop dead in the water.[6] Other Soviet vessels in Alexandria were ordered to leave port. This again removed unnecessary and vulnerable Soviet targets and gave Moscow an immediately available shipment of arms with which it could make a dramatic gesture early in the war, should the circumstances be propitious.

Fifth, the Soviet Union ensured that the SCUD brigades in Egypt remained under strict Soviet control. This weapon of deterrence, so crucial to Sadat's decision to go to war, was a key element in a strategy of escalation control. Exclusive operational possession of the weapon by the Soviet Union meant that the war could not escalate to strategic raids on population centers without Moscow's specific approval.

Finally, Soviet Ambassador in Cairo, V. Vinagradov, was given specific instructions to seek an early end to hostilities. Some early coordination in this regard was probably effected with Syrian President Hafez Assad, who shared the Soviet contempt for Egyptian military capabilities and had no affection for Sadat personally. This coordination probably resulted in an agreement between Syria and the Soviet Union to press for a cease-fire early in the conflict.[7]

One significant omission in Soviet precrisis actions was any warning of the impending attack to the United States, an action that, at least from the perspective of the United States, was required by the communique on the "Prevention of Nuclear War." Admittedly, there was a host of signals sent that the U.S. could have correctly interpreted as indicating that hostilities were imminent, including the very public evacuation of Soviet dependents, but at no time did Moscow consult with the United States on the impending outbreak of hostilities.[8]

Thus, by October 6, 1973, the Soviet Union had positioned itself to try to control and exploit a war that it would have preferred not to see occur. Having been brought to the brink by a persistent and clever client, Moscow now faced the task of managing the crisis and avoiding damage to détente.

Egyptian Objectives in the War

Throughout his presidency, Sadat had never wavered in his intention to resume hostilities against Israel and to recapture the Sinai. But Sadat was unencumbered by the anti-Israel rhetoric that had marked Nasser's reign.[9] He was prepared to live with Israel and even extend diplomatic recognition, as presented in his peace proposals of 1971. However, like his predecessor, Sadat could not allow the occupation of the Sinai to become a permanent fixture in the region. Sadat saw a renewed conflict as his only option to effect the return of the Sinai from Israeli control. This was Sadat's overriding objective, and he saw several subordinate objectives within the conflict that could serve this end.

First, Sadat, too, wanted to limit the conflict. He believed that he could not hope to defeat Israel in an extended war or to recapture the Sinai through force of arms alone. Based on 25 years of experience, the "myth of Israel's . . . superior, even invincible air force, armory, and soldiers," which Sadat says the world had accepted, had become ingrained in the Egyptian political-military psychology.[10] Sadat's military objectives were, therefore, limited in scope. He first wanted to force the Suez Canal and overrun the Bar Lev Line. Second, he wanted to press the attack and seize the key

Sinai passes that controlled access to the interior of the Sinai itself. Sadat did not envision an attack beyond those limited objectives, and he probably thought that even those objectives were not attainable.[11]

Second, Sadat sought to restore military dignity to the Arab world and Arab leadership to Egypt. This he wanted for both psychological and pragmatic reasons. Psychologically, the entire Arab world had suffered an enormous loss of pride in the 1967 war and in the continued occupation of Arab lands. There was no chance for a regional peace with Israel as long as the stigma of defeat continued to mark the Arab world. Any form of political settlement had to have, as its antecedent, some sort of Arab military victory. Pragmatically, Egypt desperately needed Arab monetary support for the failing Egyptian economy. Loss of revenue from the Suez Canal, which was closed and mined in 1967, was a crushing blow to the Egyptian economy, a blow compounded by the debt-servicing requirements imposed by the Soviet Union. Without a restoration of Arab pride under Egyptian leadership, Arab support for the Egyptian economy was highly improbable. The restoration of such pride could only be accomplished through a renewal of the war with Israel and some demonstration of success on the battlefield.

Third, Sadat wanted to bring the Middle East situation back to center stage in superpower relations. Having seen how, in two successive summits, Egyptian interests had been sacrificed on the altar of détente, Sadat knew that only through a renewal of the conflict could Egypt hope to force superpower attention on his security situation. The superpowers, Sadat realized, held the key to a political solution but were willing to see the "no war, no peace" situation continue indefinitely as long as it kept the Middle East from becoming a source of friction in an otherwise increasingly relaxed superpower relationship. Only through a resumption of the war could Sadat break this attitude.

Fourth, Sadat sought to force the United States to deal with Israeli intransigence. Only by eliminating the perception of Israeli invincibility from American minds could Sadat expect the United States to play a constructive role. Sadat anticipated that, once the

United States became aware that Egypt was not a "motionless corpse," Washington would become more interested in forcing its Israeli client to be more flexible at the bargaining table.

Taken together, Egyptian objectives at the outset of the war were not wholly incompatible with those of the Soviet Union. Both patron and client wanted to see a limited conflict with a modest military achievement by Egypt and a net setback for Israel. But this apparent compatibility masked a deeper division, one that is normal in patron-client state relationships. Sadat's most basic goal was to reduce his threat environment, both internal and external, by the use of a limited conflict to force political concessions from Israel. He saw the war as a means to make progress on the resolution of the Israeli occupation and, thereby, reduce the threats to his regime and to Egyptian security. The Soviet Union, on the other hand, did not want to see Egypt's threat environment reduced. This would cause the relationship to move from type III, in which the patron seeks goals of strategic advantage in a high-threat environment of the client, to type VI, in which the patron has the same goals but the client's threat environment is much lower. Such a shift in Egypt's threat environment would cause a substantial loss of influence in the relationship for the Soviet Union; type VI relationships are appropriately called "client centric." Worse still, reduction in Egypt's threat environment could cause the relationship to end altogether, particularly if the United States were to play a major part in resolving the Israeli occupation of the Sinai. A demise of the relationship would cause the Soviet Union to lose its coveted access to Egyptian naval facilities, which, even if somewhat reduced since 1972, was still a major strategic asset for the Soviet navy.

This basic incompatibility, when exacerbated by the pressures of the October War, would eventually result in the outcome Moscow feared most: the end of its 20-year relationship with Egypt.

THE OCTOBER WAR

The Battle Problem

The tactical problem that Sadat faced was significant, and those who tend to heap abuse on Arab military capabilities would do well to remember Egyptian accomplishments in October 1973.

The Sinai presented a difficult battle problem for Sadat. One hundred fifty miles wide and 250 miles long, this peninsula was divided topographically into four distinct areas. In the south, the terrain was broken, mountainous, and virtually impassable to any modern army. The central portion of the Sinai was substantially flat desert terrain, sloping away from the mountainous south to the Mediterranean coast. There were few defensible positions in the central Sinai, nor did the 1949 armistice line between Israel and the Sinai (the pre-1967 boundary) provide any natural defense. The third area in the Sinai was a strip of land, some 30 miles wide, along the eastern bank of the Suez Canal. The fourth, and most important tactical area was the line of high, broken hills in the western Sinai running from the Mediterranean south to the highlands. This line of hills effectively blocked the advance of armies from the east or the west. Egress through the mountains was possible only through three primary avenues of approach: the Giddi and Mitla Passes and the coast road. Because of the sand seas that bordered the coast road and severely curtailed off-road mobility, only the Giddi and Mitla Passes were really useful in attacking east or west in the Sinai. Possession of the passes was thus crucial for control of the Sinai. The role of these passes, dictated by the topography of the Sinai, was critical in determining the outcome of the October War and the dynamics of crisis management.

In order to reach the passes, however, the Egyptian Army had first to face the crossing of the Suez Canal and the storming of the Bar Lev Line. The Canal itself was a substantial obstacle made more difficult by prior Israeli defensive preparations. Moreover, Israeli had built up the eastern bank of the Canal to heights of over 47 feet, providing a commanding view of any Egyptian effort to cross the Canal and blocking Egyptian ground reconnaissance from the western bank.[12] Most significantly, Israel had established the Bar Lev Line itself, a series of mutually supporting hard points designed to defeat an Egyptian attack at the edge of the Canal or, failing that, to contain an Egyptian crossing until reserve units could be brought up from the central Sinai or from Israel.

The difficulty in executing a successful river crossing is that only a small number of forces can cross initially and then must hold the area on the hostile bank (the bridgehead) until bridges are built. This initial wave of forces cannot normally be equipped with heavy support, such as tanks or artillery, and must be prepared to repel counterattacks from a heavily armed enemy. After the initial bridges are established, they themselves are subject to attack and destruction, and the forces in the bridgehead are always susceptible to isolation and defeat. Next to an amphibious landing, a river crossing is perhaps the most difficult military operation.

With this as background, one can see that Sadat's tactical problems were by no means trivial; the undertaking of even a limited war in this environment was a courageous decision.

Saturday, October 6

At 1400 hours local time, on October 6, the armies of Egypt and Syria launched simultaneous and coordinated attacks upon Israeli defensive positions along the Suez Canal in the south and the Golan Heights in the north.[13] In both areas, Israeli forces were at substantially reduced levels, having not expected an attack and having sent large numbers of troops back to Israel for the Yom Kippur holy days.[14]

In the Golan Heights, the Syrian Army attacked with three mechanized infantry and two armored divisions numbering some 28,000 troops, 1,000–1,250 tanks and 600 artillery pieces.[15] There they met an Israeli front held by essentially two understrength armored brigades, containing about 70 tanks each, and backed up by several infantry battalions dispersed into various defensive positions. Israeli strength was, in reality, less than a fully operational brigade and was, therefore, outnumbered more than twelve to one.[16]

In the south, the Second and Third Egyptian Armies swept across the Suez Canal in an impressive display of combined arms tactics, crossing the Canal at five separate points. Because of the nature of river crossings, Egypt had to rely primarily upon its infantry divisions, leaving its tanks on the western bank to await the construction of heavy bridges. Using Soviet-made light bridges and some novel tactics,[17] the Egyptians did not pause to regroup once they were on the east bank but rather drove immediately to force the Bar Lev Line.[18]

The greatly outnumbered and surprised Israeli defenders presented only light opposition to the determined Egyptian attack that sought, in one afternoon, "to wipe out the stigma of defeat and humiliation under which the Egyptian Army had suffered for six years."[19] The infantry units of the Third Army swept south along the Canal, rolling up Israeli defensive positions with little difficulty. Israeli aircraft, which had accomplished so much in the June 1967 war, were effectively neutralized by the massed SA-2, 3, 6, and 7 missiles deployed along the west bank of the Canal. The few times that the Israeli Air Force was able to penetrate the air defense umbrella and destroy a bridge, the Egyptian crews were able to replace it quickly with another.

By 1800 hours, less than 240 minutes after the assault began, the two Egyptian armies controlled almost the entire east bank of the Canal, the Second Army in the north and the Third Army in the south. As darkness fell, the Egyptian forces established defensive positions, deploying numerous ambush patrols along likely Israeli avenues of approach from the Giddi and Mitla Passes, which controlled access to the Suez Canal area from the central Sinai. As the three Israeli brigades kept in reserve near the Canal attempted to throw the Egyptians out of their bridgeheads, they were met with a withering fire from ATGMs and small arms. Of the nearly 300 tanks within these three Israeli brigades, more than 250 were destroyed during the night.[20] The effective reduction of these units left virtually no Israeli reserve in the entire Sinai.

Thus, in the first hours of the conflict, Egypt scored an impressive tactical victory. No less substantial was the advance of the Syrian Army, which by nightfall had driven the Israeli defenders in the Golan Heights back some five miles.

In the midst of this success on the battlefield, the Soviet Union made its first major effort to control the crisis. At 1940 hours on October 6, Soviet Ambassador Vinagradov called on Sadat to ask that the Egyptians accept a cease-fire, based upon an alleged request received from Assad. Sadat was astounded that, with the Egyptian forces experiencing tactical success on the battlefield, the Soviet

Union would suggest a cease-fire. Vinagradov confirmed that his instructions came from Brezhnev. Sadat's reply was uncompromising: "I'd like to inform the Soviet leaders that, even if Syria did demand it, I won't have a cease-fire until the main targets of my battle are achieved."[21] Sadat could have better understood "if the proposal for a cease-fire had come from Washington, since the battle was going the Arab's way."[22] Sadat hastily dispatched a cable to Assad, asking for an explanation of the Soviet allegation that Syria had requested the cease-fire.

It is extremely doubtful that the Soviet Union knew, at this point, about the extent of Egypt's tactical success. There is no evidence that any Soviet advisers accompanied the Second and Third Army, and it is unlikely that real-time intelligence was reaching Moscow. Even had the Soviet Union believed everything it heard on Radio Cairo, it would not have had an accurate picture of the day's events in the war; contrary to the public reporting of most wars by a belligerent in the Third World, the reporting of the Egyptian media consistently understated the progress of the Arab forces in the beginning of the war.

The Soviet request for a cease-fire was, in all probability, based on the prewar assessment that Egypt would do very poorly and that the war should be stopped almost as soon as it began. The Vinagradov request was the implementation of one of the crisis-management tools established by Moscow prior to the outbreak of hostilities. The Soviet Union sought an early end to the war before Egypt could be decisively defeated, the Soviet commitment escalated, or the United States become involved. The unexpected success on the battlefield in the initial hours of the war did not factor into the Soviet decision to request a cease-fire and effectively thwarted this initial Soviet effort at crisis management.

Sunday, October 7

By the morning of the second day of the war, the two Egyptian armies on the east bank of the Suez Canal had consolidated their bridgeheads and began moving many more SA-6 batteries across the Canal.

The Egyptian forces found themselves with only sporadic opposition throughout the day. The Israeli Air Force was ordered to stay away from the Canal in order to avoid further losses to the SAMs positioned all along the Canal and because their presence was in far greater demand in the north. Egypt found itself in the unaccustomed position of not knowing how to exploit a tactical victory.

Far more intense action took place on October 7 in the Golan Heights. Because of the proximity of the Golan Heights to Israeli population centers, Israel sent the bulk of its forces north to engage the Syrians. Israeli strategy became to concentrate its initial efforts against Syrian forces in the Heights, drive them back, and then shift their forces to the south to deal with Egypt. This strategy allowed for greater efficiency for all Israeli forces but especially the Air Force.

The Israeli strategy proved to be justified as Syrian armor continued to advance in the Golan Heights, although at an alarming cost in terms of tanks and soldiers; Israeli forces achieved kill ratios of some twelve to one.[23] By nightfall on October 7, the Israeli position in the Heights was desperate, and at one point, Syrian forces penetrated far enough into the Heights that they could see the Hula Valley below. The tactical question hinged on whether or not Israel could commit sufficient forces to the battle to stem the Syrian advance before the Syrians succeeded in overrunning the Heights entirely.

From a collective Arab standpoint, the night of October 7 represented the high watermark of the war. Once again, in the midst of this tactical success, Ambassador Vinagradov materialized at Sadat's residence in order to again request that a cease-fire be announced. He repeated his assertion that Assad had requested such a cease-fire, this time due to mounting Syrian tank losses in the Golan Heights. Sadat, however, had received an answer to his cable in which he asked Assad to confirm that he had, indeed, requested a cease-fire. Assad denied that he made such a request. Sadat answered Vinagradov's second request for a cease-fire by saying:

> Now listen, this subject is closed; I don't want you to take it up any further with me. You know, and I told you yesterday, that I won't have a cease-fire until the objectives of this battle have been achieved. I'd like you to tell the Moscow leadership to send me some tanks at once. This will be the biggest tank battle in history.[24]

This second request for a cease-fire was, in all likelihood, a continuation of the execution of Vinagradov's precrisis instructions to effect a cease-fire early in the conflict. Although some of the public pronouncements in the Israeli press were admitting the breach of the Bar Lev Line, Moscow still had little solid information on which to base a general reassessment of its prewar planning. Moreover, even had Moscow known the details of the Arab success, it would have had little reason to change its political tactics. The acceptance of a cease-fire at this point would have left both Egypt and Syria with substantial tactical gains but nowhere near the decisive victory that Egypt would need to assuage its high-threat environment.[25]

Vinagradov was not without some flexibility in dealing with Sadat on October 7. When it became apparent that Sadat was not going to accept a cease-fire, the Soviet ambassador was authorized to tell Sadat that the Soviet Union intended to establish an air bridge over which "long overdue ammunition and equipment would be airlifted to Egypt immediately."[26] The same day, Soviet merchant vessels began to leave Odessa in the Black Sea with military supplies for both Arab clients.

Once again, these acts were probably part of the preplanned tools of crisis management established by the Soviet Union before the outbreak of the war. The promises extended to Sadat by Vinagradov did not, in themselves, reduce Soviet flexibility or expand the Soviet commitment to its Arab clients. Indeed, the dispatch of merchant vessels and the promise of an air bridge did not actually result in any immediate deliveries. It is probable that the Soviet Union anticipated that its promises, coupled with the apparent Arab success on the battlefield, would entice Sadat into accepting a cease-fire. In the intense environment of a crisis, this represented a contraction of the time-honored Soviet technique of trying to manage its relationship with Egypt through the use of promises, rather than deliveries. In any event, the Soviet Union wanted to be in a position to refute any claims that it had shirked its responsibilities to its clients, claims that Sadat was sure to make if the conflict began to turn against the Arabs and the Soviet Union did not make immediate equipment deliveries.

At the same time, the Soviet Union sought to insulate détente from the effects of the war. On October 7, Brezhnev and Nixon exchanged messages, vowing to ensure that the war did not seriously damage U.S.-Soviet relations or undermine the progress that had been made over the past three years and two major annual summits between the leaders of the superpowers.[27] In an effort to further downplay the significance of the Middle East war to superpower relations, Nixon did not even use the Washington-Moscow "hot line" to communicate with the Soviet leadership.[28]

In this effort, the Soviet Union wanted to ensure that, should the war begin to get out of hand, the Soviet Union would have already established with the United States the primacy of détente. With this political flank more firmly anchored with the United States, the Soviet Union felt better able to deal with the crisis at the client level and to exploit opportunities that might present themselves in the course of the conflict.

Monday, October 8

On the Sinai front, the two Egyptian armies continued to consolidate their hold on the east bank of the Suez Canal and extended their bridgeheads north and south. But, as during the previous two days, Egyptian forces failed to exploit their tactical advantage and seize the passes.

In the Golan Heights, the battle continued with a ferocious intensity. By October 8, Israel had completed a full mobilization of its reserve forces and began to shift to the offensive in the north. After two days of intense combat, the Syrian capacity for offensive military operations was successfully eliminated.

Losses on both sides were extremely heavy during the Golan phase of the war. Syria had nearly 1000 tanks destroyed and lost some 6000 military personnel by the end of October 8.[29] Israeli losses had also been substantial, including some 200 tanks and some 500 personnel.[30]

On the same day, U.S. Secretary of State Kissinger, as interested as his Soviet counterparts in insulating détente from the vagaries of the battlefield, spoke at a *Pacem in Terris* gathering in Washington where he asserted that détente was the highest priority of U.S. foreign policy but warned that détente could not survive "irresponsibility in the Middle East."[31] To Moscow, this meant that détente could not be completely isolated from the conflict and that, therefore, some measure of caution was still in order. Kissinger was deliberately ambiguous on his definition of "irresponsibility," both to provide the U.S. with flexibility in its own reactions to the conflict and to avoid laying down an irrevocable marker on Soviet behavior.

In Cairo, Vinagradov repeated again his request for a cease-fire and was again rebuffed by Sadat.[32] In fact, in retrospect, October 8 would have been an ideal time for the Arabs to accept a cease-fire; all along both fronts, the Arab armies had made substantial gains and had recovered occupied Arab lands. Israel had not yet mounted serious counteroffensives in either area.

However, in this case, as in most crises, accurate information was a most precious, and scarce, commodity. No principal leader in the crisis really knew what the true situation was on the battlefield at any given moment; decisions were often made on the basis of tactical situations fully hours or even days old. For example, on the evening of October 8, after Syria had lost nearly 1000 tanks in battle, Assad still reported to Sadat that

> the battle was going well, as far as Syria was concerned. They were inflicting heavy losses on the enemy and had already liberated more than half of the Golan Heights. Their losses were not abnormal and could be replaced from their own reserves.[33]

The superpowers were plagued by a similar lack of information, a deficiency that grew as the battle became more complex.[34] It comes as no great surprise, therefore, that the Soviet Union surfaced initiatives that appeared to be dated, especially after the list of prepositioned crisis-management measures had been exhausted.

Tuesday, October 9

Israel had generally regained control of the Heights by October 9 and had pushed the Syrian Army behind the "purple line," the line at which the war had begun three days earlier.

On the Sinai front, the battlefield remained generally quiet. The two army bridgeheads were each six to ten miles deep, the Bar-Lev Line had been almost completely reduced, and "the Egyptian Army and the Egyptian people were enjoying their finest hour since the days of Suez in 1956."[35]

Having accomplished the first of Sadat's two operational objectives, Egypt now adopted an "operational pause," in accordance with what Sadat claims was the prewar plan with Syria. Egyptian caution at this point, however, was tactical suicide. Heikel says:

> The spectre of previous defeats inhibited those in command from taking anything that could possibly be construed as a risk. The security of the Army may have weighed more with them than the exploitation of an unlooked for degree of success. It is my belief that had the passes been reached and occupied, the whole of the Sinai would have been liberated with the incalculable political consequences which would have flowed from such a victory.[36]

Heikel's explanation of the failure of the Egyptian Army to seize the Sinai passes is compelling. Certainly, had the Egyptian forces seized and controlled the passes, Israel's battle problem would have been significantly compounded and could well have precluded the tactical reverses that Egypt suffered later in the war.

On October 9, the list of Soviet preplanned crisis-management initiatives was largely exhausted. Over the first three days of the war, Moscow had succeeded in obtaining general statements from the United States to maintain détente despite the Middle East war, but the Soviet Union had succeeded in little else. It had not been successful in obtaining a cease-fire, nor had it controlled the level of the conflict. The Soviet Union now began to shift its crisis-management efforts to exploit the wholly unexpected success of its Arab clients. Adding further impetus to this shift was the feeling of relative security that Moscow had as far as the conduct of the crisis at the superpower level was concerned. The Kissinger statement of

October 8 and the various contacts between Nixon and Brezhnev had convinced the Soviet Union that the United States would be willing to tolerate some form of increased Soviet involvement without placing détente at risk. To be sure, Soviet involvement would have to proceed cautiously in order to ensure that an unexpressed American level of tolerance would not be violated. But, all this considered, October 9 marked a departure for the Soviet Union from its erstwhile low-key posture on the war.

The mechanism for signaling this shift was the Brezhnev response to an October 8 letter from Algeria's President Boumedienne. In the response, Brezhnev urged Arab leaders to "use all means at their disposal and take all required steps with a view toward supporting Syria and Egypt in the difficult struggle. . . . Syria and Egypt must not be alone in the struggle."[37] Brezhnev was careful to avoid any promise of direct Soviet involvement; he instead urged Arab support for the Arab struggle "against a treacherous enemy."[38] Moscow hoped that, by serving as a rallying point for the Arab world, its postwar stature would be greatly enhanced. Sadat thought that "they [the Soviets] saw the situation moving in a very favorable direction and felt this was their chance to regain most or all of their lost prestige in the Middle East."[39]

However indirect, the Brezhnev letter was the first escalation of Soviet involvement in the war and the beginnings of the collapse of Soviet crisis management. Up until that point, Moscow had used its series of preplanned measures designed to control the situation; after October 9, the Soviet Union was increasingly swept along by events over which it had a decreasing amount of control. This phenomenon, endemic in patron-client state relationships, was to have major implications for superpower relations.

At the same time, the United States was making a decision to increase its own involvement in the war. Richard Nixon remembers that

> by Tuesday, October 9, the fourth day of the war, we could see that if the Israelis were to continue fighting, we would have to provide them with planes and ammunition to replace their early losses. I had absolutely no doubt or hesitation about what we must do. I met with Kissinger and told him to let the Israelis know that we would replace all their losses.[40]

Thus, on October 9, both superpowers began to increase their involvement in the client crisis, but they each retained their intentions to keep the crisis out of the direct superpower arena. Both the Brezhnev letter and the Nixon commitment to airlift to Israel were taken without consultations and without a good deal of thought being given to the reaction of the other superpower.

Wednesday-Thursday, October 10-11

The Golan front continued to deteriorate for the Syrian Army, as Israel methodically drove out of the Heights and onto the plain of Damascus. The Israeli commanders decided, on October 10, not to be content with simply restoring the "purple line," but rather, they resolved to destroy Syrian war-making potential altogether before turning to the situation in the Sinai. Israel also embarked on a substantial escalation of the war on the Syrian front; the Israeli Air Force conducted strategic bombing missions on the city of Damascus, destroying among other things the Soviet cultural center and killing a Soviet national civilian employed there.

On the Sinai front, the situation remained stable. But now, the Soviet Union began to exert subtle pressure on Sadat to move forward out of the bridgeheads to seize the passes, both for strategic reasons and to relieve some of the tactical pressure on the Syrians.[41] Sadat, however, was not to be budged.

In response to these developments, the Soviet Union now took a more active, direct role in the conflict. After the bombing of the Soviet cultural center in Damascus, Moscow laid out stern warnings to Israel that "grave consequences could result from Soviet casualties."[42] Moreover, the Soviet Union made it clear that it was unwilling to stand by and watch Israel conquer Damascus or completely reduce the Syrian Army. Such attempts would result in direct Soviet military intervention.[43] These were the first threats of Soviet intervention in the war, and, by making such announcements, the Soviet Union was already placing limits on its options and making them directly dependent on the actions of the participants in the client crisis.

The Soviet Union escalated its involvement in yet another way. In response to the deteriorating Syrian tactical position and the massive losses of equipment over the first four days of the war, the Soviet Union began its promised airlift to Syria on October 10 and to Egypt the following day. This equipment was drawn from prepositioned war reserves in Eastern Europe that were readily available for rapid movement. Heavier equipment, such as main battle tanks, had already been prepared and embarked in ships at Odessa. Although Moscow had anticipated the requirement to resupply its Arab clients and had prepositioned munitions and equipment to expedite the process, the Soviet leadership would have undoubtedly preferred to begin such an effort after hostilities had ceased when, as in 1967, maximum advantage could have been extracted with minimum risk. But the nature of the battle provided no such luxury.

Also on October 10, the Soviet Union placed three of its airborne divisions on alert, raising the specter of direct Soviet involvement in the conflict and putting teeth into Soviet threats directed against Israel.[44] In all probability, the latter objective was more relevant to the Soviet alert; Moscow knew that the alert of its airborne divisions would be detected by the United States and promptly passed along to Israel.

Thus, by Thursday, October 11, the Soviet Union found itself increasingly enmeshed in the client crisis. Crisis management had turned from control of the situation and limiting the conflict through an early cease-fire to direct involvement in the perpetuation and extension of the conflict.

However, on the second level of crisis management, that of managing relations with the United States, the Soviet Union still attempted to avoid escalation. Consultations between the superpowers continued, and on October 11, the Soviet Union agreed on Egypt's behalf to an October 10 Kissinger proposal for a cease-fire in place.

The apparent contradiction between the beginning of the Soviet airlift, a move guaranteed to keep the war going, and Soviet-U.S. agreement to a cease-fire on behalf of their respective clients is explainable in the context of patron-client crisis management. The patron will often find itself having to deal with conflicting objectives at different crisis levels. The contradictory Soviet behavior on October 10–11 was due to Soviet intentions to maintain its credibility with its Arab clients while placating United States desires to end the conflict. Although this appeared to work on October 10–11, the irreconcilable nature of Soviet actions would soon force Moscow to establish a clear priority in its crisis-management efforts at both levels.

Friday, October 12

The Israeli counterattack in the Golan Heights continued to push Syrian forces toward Damascus. Particularly important in this regard was Israeli control over the Kuneitra-Damascus highway, a high-speed avenue of approach into the Syrian interior.

On the Sinai front, Egypt began at last to prepare to continue its offensive eastward. Egyptian armor, hithertofore withheld on the western bank of the Suez Canal was now committed to the bridgeheads on the east bank.[45] Additional SA-6 batteries were deployed into the bridgehead in order to extend the air defense umbrella.

The Soviet airlift to both states continued, although still at a relatively modest rate. Moscow was essentially waiting for a signal from the United States as to whether the crisis at the superpower level would intensify as a result of the Soviet airlift. Heavy equipment, such as tanks and howitzers, of the type that Sadat had requested from Vinagradov on the second day of the war, had not yet materialized. In fact, only Yugoslavian President Tito had made any firm commitment to deliver tanks to Egypt.[46]

The answer from the United States helped relieve the Soviet crisis-management contradictions, at least for the moment. The new Secretary of State, Henry Kissinger, in his first press conference, announced that, although Soviet actions had not been "helpful" in controlling the client crisis, the United States did not "consider the Soviet actions as of now constitute the irresponsibility [cited in the October 8 speech] that would threaten détente." Kissinger went on to say:

> If you compare their conduct in this crisis to their conduct in 1967, one has to say that Soviet behavior has been less provocative, less incendiary, and less geared to military threats than in the previous crisis.[47]

This, of course, should have come as no great shock; the 1967 war had begun largely as a manipulated crisis by the Soviet Union in an effort to secure greater goals of strategic advantage. The 1973 war was, by contrast, not a manipulated crisis, and the Soviet Union, like all participants, had to play a more cautious, restrained role until the situation had become sufficiently clarified so that Moscow could determine which course of action it should pursue.

From the Soviet perspective, the Kissinger press conference meant that the United States was prepared to decouple the two crisis-management levels and that Soviet rhetoric and the rather tentative airlift that marked October 9, 10, and 11 were acceptable behavior in the context of détente.

This was not to say that the United States was prepared to see its client lose on the battlefield. Although it had reversed earlier setbacks in the Golan Heights, Israel had itself experienced substantial losses and needed a resupply effort. As mentioned, Nixon had already made a decision to airlift equipment to Israel, but the implementation of this airlift got off to a slow start. The president and the Department of State favored an immediate airlift beginning on October 9. The Defense Department, fully aware of the implications such an effort would have on U.S. readiness, balked and stalled. The issue was finally resolved on October 12 when Nixon, in a fit of temper, ordered Schlesinger to comply with his orders. He remembers saying, "Goddamn it, use everyone [of the airplanes] we have. Tell them to send everything that can fly."[48]

At this point, then, both patrons were willing to focus their mounting involvement and their crisis-management efforts on the war itself and implicitly agreed to keep the bilateral superpower crisis at a low level.[49] Both demonstrated their willingness to tolerate expanding involvement in the client crisis by the other.

With Moscow's crisis management efforts at keeping the level-two crisis with the United States at an acceptable intensity, the Soviet Union began to increase once again its involvement in the conflict. On October 12, Israeli missile boats sank the Soviet freighter Ilya Menchikov in Tartous Harbor, Syria. This elicited yet another ominous Soviet threat. *Tass* warned that:

> If the ruling circles of Israel assume that their activities in regard to peaceful cities and civilian objects of Syria and Egypt will remain unpunished, then they are deeply mistaken. . . . The Soviet Union cannot remain indifferenct to the criminal acts of the Israeli military.[50]

These warnings, coupled with the Soviet airlift that accelerated almost as soon as Kissinger's news conference ended, signaled the deepening Soviet involvement in the war and Soviet intentions to ensure that the outcome was favorable to its Arab clients.

Saturday, October 13

Bouyed by the Soviet airlift and the arrival of major end items by sea, as well as by the entry of Iraqi, Jordanian, and Moroccan troops into the war, the Syrian front began to stabilize near the town of Sassa. As the Israeli Army advanced, it was met with an increasing volume of fire, until one Israeli officer commented that advancing Israeli units "met with an avalanche of people and tanks."[51] By nightfall, the Israeli forces were near Sassa on the Damascus-Amman highway and were within artillery range of Damascus. The Israeli Army, exhausted from the slow, deliberate advance under intense Syrian pressure, settled into the best defensive positions they could find. The Soviet threat of intervention, forces from other Arab states, Soviet resupplies for Syria, and the potential for high casualties all combined to stop the Israeli offensive. Moreover, up to this point, the Jordan front had been quiet. Neither Israel nor Jordan was anxious for fighting to begin along the West Bank, and cutting the Amman highway would have put Hussein in a position where he would have had to retaliate there.[52] In addition, and of primary importance, Israel still faced a formidable foe in the Sinai. These factors combined effectively to bring the Golan phase of the October War to an end. Arab forces would launch a number of small-scale counterattacks during the balance of the war but these would achieve very little, and the main focus of operations, after October 13, shifted to the Sinai.

The Sinai front remained essentially static throughout October 13, although in response to Assad's increasingly strident pleas, Egyptian forces were gathering in the bridgeheads for an attack the next day.[53] Sadat felt that, with Assad apparently now in trouble, Egypt was called upon to relieve the pressure on its Syrian ally by an attack against Israel's southern flank. To support this pending attack, Egypt moved more of its armor across the Canal and into the bridgeheads on October 13.

Even as these preparations for an expanded conflict were under way, the efforts by the superpowers to obtain a cease-fire continued unabated. On the morning of October 13, the British ambassador to Egypt appeared, on behalf of the United States, with whom Egypt still had no diplomatic relations, to ask if Sadat would confirm his acceptance of the Kissinger cease-fire that had been agreed to by the Soviet Union speaking for its client. Sadat angrily rejected this appeal and was especially annoyed that the Soviet Union would presume to speak for Egypt on matters of this import. Through Presidential Adviser Hafez Ismail, Sadat had already told Kissinger in 1972 that the Soviet Union did not speak for Egypt and that the United States should deal with Egypt as a sovereign power. Sadat's answer to the British ambassador was unambiguous:

> I haven't agreed to a cease-fire proposed by the Soviet Union or any other party. He [Kissinger] should contact Cairo, not Moscow, in respect of anything concerning Egypt. Furthermore, I shall not agree to a cease-fire until the tasks included in the plan have been accomplished.[54]

This indeed presented a substantial problem for Soviet crisis management. Moscow assumed that the United States had almost complete control over its client, Israel. The demonstrated independence of Cairo from Soviet control now meant that the United States could deal directly with the crisis at both levels, whereas Moscow only had control over the crisis at the superpower level. The Soviet Union could not now expect to be able to effect a cease-fire on any terms except Sadat's. By contrast, the United States, in Soviet eyes, could tell its client exactly what to do and when to accept a cease-fire. The rejection of the Kissinger initiative by Sadat and the accompanying harsh words directed against the crisis-management efforts of the Soviet Union were signal defeats for Soviet efforts to control the client crisis. On October 13, therefore, Moscow found itself only with the option to continue to expand its commitment, to match that of the United States, in order to ensure that its clients did not suffer another humiliation at the hands of the U.S. client.

This realization was further stimulated by the beginnings of the U.S. airlift to Israel. In the initial stages of the airlift, most U.S. flights originated in Germany, where prepositioned stocks of U.S. equipment were available for immediate movement. The use of these stocks by the United States to resupply Israel was a reflection of the confidence that both superpowers shared that the situation could be contained to the level of a client crisis and not be allowed to spill over into the level two of détente. Both the United States and the Soviet Union drew down stocks of equipment from the Central Front in Europe, equipment that would be vitally needed in the event of a general war between the United States and the Soviet Union. By using that equipment to resupply their clients, both superpowers demonstrated the low level of crisis they perceived between themselves. The NATO allies of the United States did not share this optimism and began a vigorous protest to Washington over the use of European-based equipment to resupply Israel.[55]

Somewhat paradoxically, the fleets of the two navies began a concerted buildup in the Mediterranean on October 13. Although the U.S. Sixth Fleet far outclassed the Soviet Mediterranean Flotilla, the Soviet Navy was, for the first time, presenting a credible threat to major U.S. combatants.[56] Although both fleets were taking pains to stay clear of each other, it was evident that, should the situation in the client crisis so require, a major sea battle between the superpowers could quickly erupt.

This naval buildup is explainable when seen from the perspective of the client crisis. Should either side find it expedient to intervene in the war, naval power was the most readily available asset that could be brought to bear. It is doubtful that, at this point, the respective navies were in the region in anticipation of direct conflict. Rather, they were there to support any required intervention in the client crisis. This point is reinforced by the presence in the Soviet Flotilla of naval infantry. Although of limited tactical effectiveness, due to their lack of organic firepower or logistical support, the naval infantry forces provided the Soviet Union with the capability to insert forces rapidly for symbolic or deterrence purposes in the client crisis.

Whatever the underlying purposes behind these naval buildups, the presence of large naval formations in the relatively narrow confines of the eastern Mediterranean provided ample opportunity for direct superpower conflict, planned or otherwise.

Sunday-Monday, October 14-15

After nearly a week of a static defense, the Sinai front erupted into the largest tank battle in history as the Egyptian Second and Third armies struck out of their bridgeheads towards the Sinai passes. From there, the Egyptian forces, led by their armor spearheads, were to be prepared to attack eastward.

However, what was readily attainable on October 6–7, was impossible one week later. Once outside of their bridgeheads, the Egyptian columns lost their air defense umbrella and were exposed to the combined arms tactics for which Israel was so famous. The respite of a week had provided Israel with the opportunity to dispatch reserves to the Sinai, and the effective end of the battle on the Golan front the day before freed the Israeli Air Force to engage the Egypt units once they emerged from underneath the protection of their SAM batteries.

The Egyptian attack faltered, although not until after a valiant fight, and was finally repulsed with apalling losses.[57] In an effort to prevent an Israeli counterattack, and in order to replace some of his tank losses, General Ahmed Ismail committed his stategic reserve to the east bank. The armor heavy First Egyptian Army moved across late in the day on October 14 and throughout the next day to bolster the decimated armor forces of Second and Third armies. Along with the failure to seize the Sinai passes, this decision was the most significant strategic blunder committed by Egypt in what was otherwise an impressive military performance. Ismail violated one of the principal maxims of war: "never commit the reserve to redeem a failure," and in so doing, set the stage for the military disaster that was to befall Egypt in the closing days of the war.

Increased U.S. involvement, in the form of the airlift, provided Israel with the lift it needed and, by nightfall on October 14, Egypt had returned to its positions in the two bridgeheads. Ismail now decided that he would establish effective defensive positions and "allow Israel to beat their heads against the wall of fire of the bridgeheads," an occupation of which the Egyptians felt Israel would soon tire.[58]

With the failure of the Egyptian offensive and the impact of the U.S. airlift, the offensive momentum swung to Israel in the Sinai for the first time. The Egyptian defensive position along the east bank of the Canal presented Israel with a challenge and an opportunity. The Egyptian armies occupied discrete bridgeheads, rather than a single, integrated front. A gap of some 25 miles separated the bridgeheads, and almost at the center of this gap, the last Israeli strong point in the center of the breached Bar-Lev Line still stood.

Israel could not hope to defeat the Egyptian bridgeheads in a head-on engagement; such an attack had been mounted on October 7 and 8 with disastrous results, and at that point, Egyptian forces were weaker than they were on October 15. Indeed, the entire Egyptian strategic reserve had been committed to the east bank, adding further difficulties to an Israel plan to reduce the bridgeheads.

But the same situation presented Israel with a risky opportunity. Because of the Egyptian force dispositions, it became feasible for Israel to drive a wedge between the Egyptian armies, force the Canal, and exploit Egyptian vulnerabilities in the now undefended Egyptian interior. Moreover, such an operation would enable Israel to roll up the SA-3 and 6 sites along the Canal and destroy the entrenched Sagger ATGM positions that commanded the approaches to the Canal.[59] The risks of such an operation were centered on the slender line of communication (LOC) that the attacking Israeli forces would have and the strong possibilities that the Israeli LOC would be cut by attacks from either Second or Third armies on the east bank of the Canal.

Because Egypt, over the long term, had greater sustainability in the war, Israel decided to take the chance and press for an early military victory. Thus, late at night on October 15, Israel attacked the gap in the Egyptian lines. By 0730 on October 16, two Israeli brigades had successfully crossed the Canal into Egypt.

Of course, battlefield intelligence being what it was, the Soviet Union knew virtually nothing of the Israeli attack, nor for that matter did the Egyptians. It is doubtful that Moscow even knew of the full extent of the Egyptian defeat on October 14–15. What Moscow did know was that Sadat had rejected the Kissinger cease-fire proposal on October 13 and that a diplomatic solution to the crisis was as far away as ever. The Soviet Union was also becoming aware of the magnitude of the U.S. airlift and of the implications this had for the ultimate outcome on the battlefield. The U.S. airlift, once underway, was running at approximately twice the daily tonnage of the Soviet airlift, a balance that Moscow knew would have an adverse long-term impact on the outcome of the war. In this environment, the Politburo decided to make another effort at a cease-fire.

To cover its diplomatic tracks in the Arab world, the Soviet effort began with some harsh rhetoric. On October 15, Brezhnev urged more support for its Arab clients in a joint communique with Algeria's Boumedienne. The Soviet Union would "contribute in every way to the liberation of all Arab territories occupied by Israel."[60] However, the translation of this rhetoric into tangible action was rather modest. The Soviet Union demanded hard cash payments for new weapons systems, whereupon Boumedienne pulled out his checkbook and provided $200 million.[61] The more serious Soviet effort, however, was focused upon the impending visit to Cairo by Kosygin.

The crisis at the superpower level on these days continued at a moderate level, although the naval forces of both superpowers continued to expand in the eastern Mediterranean. By the end of October 15, the Soviet Navy had approximately seventy vessels in the region, including the heaviest Soviet warship afloat, the helicopter carrier *Moskva*. The United States, with two carrier battlegroups in the Mediterranean already, moved a third to Gibraltar.

But, although both superpowers were willing to expand their commitments to their respective clients and to engage in some posturing in the Mediterranean, they were as yet unwilling to risk détente in a major bilateral crisis. Soviet crisis management, at this point, was a failure in managing the client crisis but successful at the patron level.

Tuesday-Friday, October 16-19

Israel continued to expand its bridgehead at Deversoir on the west bank of the Suez Canal, crossing an entire division by early morning on October 16. Not surprisingly, given the crisis environment, the lack of real-time battlefield intelligence, and Sadat's style of decentralized command, Sadat was unaware of the magnitude of the Israeli crossing throughout the entire day.[62]

Ismail assured Sadat that only "three infiltrating tanks" had crossed the Canal and that there was no cause for alarm. Of course, these three tanks were in reality an entire division expanding and improving its bridgehead virtually unmolested. Heikel argues that "this failure in communication was the biggest mistake Egypt made in the war."[63]

In this deteriorating, but as yet still ill-defined, tactical situation, Kosygin arrived at 1700 on October 16 to try to convince Sadat to accept a cease-fire. Kosygin was probably as ignorant of the tactical situation at Deversoir as Sadat was and so argued with Sadat on general terms. He proposed a four-point peace plan designed for Sadat's approval. The plan first called for a cease-fire in place, thereby guaranteeing continued Egyptian domination of the Suez Canal (bearing in mind that neither leader knew of the Deversoir bridgehead). Second, it called for eventual Israeli withdrawal from occupied Arab lands. Third, Kosygin called for an immediate international conference to resolve the Middle East conflict. Finally, the plan called for a joint United States-Soviet Union guarantee of the cease-fire.[64] To support this, Kosygin suggested that, if the United States were unwilling to commit its forces, Egypt should invite the Soviet Union to intervene unilaterally.

Sadat was unresponsive to this suggestion and went on to accuse the Soviet Union of a lack of support and of failing to provide the tanks he had requested in the opening days of the war.

This meeting reflected the basic incompatibility of the Soviet and Egyptian positions. Sadat, still unaware that the tactical fate of the Egyptian army and of the war itself was being written in the ever-enlarging Deversoir pocket, had no intention of agreeing to any cease-fire until he had secured additional territory on the east bank of the Canal. He was still confident that he could seize the passes with one additional push. Moreover, Sadat was not at all grateful for the support the Soviet Union had rendered to date. When, after the intial exchange of arguments, Kosygin lashed out at Sadat, accusing him of being ungrateful and of being a difficult ally to support, Sadat responded by saying:

> The equipment you have supplied us with is not up to date; you made us lag behind Israel in armament by a long way, and still I proceeded to fight, and—here we are. I am winning! What sort of friendly relations would you call this? Isn't it high time we buried the past and opened a new chapter?[65]

As a result of General Ismail's failure to understand the scope of the Israeli attack, only small and uncoordinated counterattacks were launched against the Deversoir pocket on October 16 and 17. These were unsuccessful, and by October 18, two Israeli armor divisions were across the Canal with 300 tanks and 15,000 men. These units began to move north and south to invest the cities of Ismailia and Suez respectively and cut the LOCs that ran through them. As the Israeli forces moved along the Canal, they overran SA-3 and SA-6 batteries, creating a gap in the Egyptian SAM coverage of the Canal. Into this gap came the Israeli Air Force to provide close air support for the Deversoir pocket and the advancing Israeli forces.

Used little in the early part of the war, the Egyptian Air Force now entered the fray. The commitment of the Egyptian Air Foce precipitated the largest air battles in the war, with substantial losses incurred by both sides. Israel emerged from these battles as the clear victor but not without having suffered heavy losses itself. Once again, the American airlift proved to make a substantial difference, for the United States provided Israel with replacement aircraft as the war was going on, while the Soviet Union provided none to Egypt.[66]

Finally, by October 18, Sadat began to recognize the size of the Israel concentration in the Deversoir pocket and the implications of the mounting Israeli drives against the LOCs of the Egyptian armies on the east bank. Ismail ordered serious and coordinated attacks against the Israeli units; reserve units outside of Cairo were sent to contain the Israeli advance, and Egyptian artillery across the Canal was turned around to fire on Israeli forces in Egypt.

Fierce fighting and high casualties on both sides continued. The Israeli Air Foce launched an all-out effort to support the ground forces, resulting in the destruction of a significant number of Egyptian tanks. This was by no means an uncontested fight; coordinated Egyptian counterattacks on several occasions blunted the Israeli advance and threatened to penetrate the Deversoir pocket.

On the east bank of the Canal, the Egyptian Third Army, which had borne the brunt of the fighting on October 14-15, could not mount an effective challenge to the Israeli LOC. Bottled up by Israeli forces surrounding its bridgehead on the east bank and increasingly threatened in its rear by the Israeli forces from the Deversoir pocket, the Third Army could do little but hope for assistance from the Second Army in the north.

On October 18 and 19, the Second Army attacked out of its bridgehead to contest the Israeli LOC. The ensuing battle was one of the costliest of the war and rivaled the 1944 Battle of Kursk in the Soviet Union for intensity of armor engagements. Once again, however, the Egyptian forces were unable to cope with the Israeli combined arms team and were forced to withdraw back into their bridgehead with substantial losses.[67] Thus both Egyptian armies on the east bank of the Canal were effectively out of the conflict and proved to be unable to challenge Israeli operations further.

By October 18, Kosygin, still having little success with his efforts to secure an agreement to a cease-fire from Sadat, was provided with satellite imagery, several days old, of the Deversoir pocket that showed the extent of the Israel threat. He hastened to Sadat in anticipation that, when shown the evidence of the impending tactical disaster, Sadat would agree to a cease-fire. He again found

Sadat unmoved by Soviet arguments, even to the extent of dismissing the Soviet pictoral evidence. Sadat still did not perceive a significant threat to Egypt and would not agree to a cease-fire.[68]

Kosygin, frustrated and unable to secure his objectives, left Cairo on October 19 with no agreement by Egypt to a cease-fire.[69] Later in the day, Sadat changed his mind and informed Moscow that he decided to accept a cease-fire, not because of Soviet pressure but rather because,

> for the previous ten days, I had been fighting entirely alone against the Americans with their modern weapons, most of which had not ever been used before. I faced the United States and Israel; while the Soviet Union stood behind me ready to stab me in the back if I lost 85 or 90 percent of my arms, just as in 1967.[70]

The timing of Sadat's acceptance and his reasoning were of critical importance in the Soviet-Egyptian relationship. Although Kosygin had worked assiduously over the previous three days to obtain Sadat's agreement to a cease-fire, it was due to American action, not Soviet, that Egypt finally accepted. The first irreparable cracks in the Soviet-Egyptian patron-client state relationship began to appear with Sadat's acceptance of the cease-fire.

That acceptance also broke down the barriers that the superpowers had erected between the levels of the crisis. Sadat's demands that the superpowers guarantee the cease-fire effectively placed responsibility for the resolution of the client crisis squarely in the laps of the patrons. It was now up to the superpowers to effect a solution.

Upon receipt of Sadat's acceptance, Brezhnev sent off a hurried cable to Nixon requesting that Kissinger be sent immediately to Moscow to work out a settlement. The letter went on to reaffirm that neither the United States nor the Soviet Union wanted to see détente harmed and that the situation in the Middle East was becoming more and more unmanageable.[71]

By October 19, the Soviet management of the crisis at both levels had experienced a curious reversal of fortune. Up to October 16, Moscow had been reasonably successful at keeping détente from becoming embroiled in the client crisis but had been unsuccessful in effecting a cease-fire and managing the client crisis itself. But as of October 19, though the cease-fire was accepted, reducing the client crisis, both superpowerss were now inextricably involved in the crisis at the superpower level.

Saturday-Monday October 20-22

Although Sadat instructed Vinagradov to cable Egypt's acceptance of the cease-fire to Moscow on October 19, the fighting continued to rage along the west bank of the Canal. By October 22, Israeli units had reached the suburbs of Suez, through which the LOCs of Third Army ran. However, the Israeli position remained tenuous on the west bank, despite its success against the LOCs of the Third Army. Sadat scraped together two additional divisions and committed them to contain the Deversoir pocket. One division in particular acquitted itself well and was responsible for preventing any Israeli breakout from the Deversoir pocket to the west.

In this tactical environment, Sadat decided not to withdraw any elements of either the Second or Third armies from the east bank of the Canal, even though their LOCs were severely threatened by the Israeli forces on the west bank. Sadat was concerned that such an action would eliminate the victory Egypt had achieved in the first few days of the war and would greatly reduce Sadat's postwar bargaining power. Sadat's military objectives were thus to keep his LOCs open to his armies on the east bank, contain the Deversoir pocket, and "take a chance and see what Kissinger [could] produce in Moscow."[72]

Both superpowers were now more eager than ever to see the client crisis end. The Soviet Union felt that any prolonging of the war would damage its Arab clients further and could well result in a direct military confrontation with the United States. The unexpected disarray within NATO that resulted from the U.S. support for Israel during the war was an achievement that needed to be exploited, but any confrontation with the United States would tend to drive NATO back together, especially if the superpower confrontation held out any possibility of escalating to a European war.

The United States, faced with a rebellion in the ranks of NATO as a result of the war and a possible oil embargo as a result of OPEC decisions on October 17 and 19, needed to end the war and begin to repair the damage that had been brought by the massive support rendered to Israel.

Thus, on October 20, in six hours of intensive talks between Brezhnev and Kissinger, the Soviet Union and the United States hammered out the terms of a U.N. Security Council resolution to stop the war. On Sunday evening, October 21, delegates and staff members of the Security Council hastily assembled to hear the terms of the proposed resolution and to vote on it.

UNSC Resolution 338, as the joint Soviet-U.S. resolution became, called for an immediate cease-fire in place on both the Sinai and Golan fronts. It reaffirmed the proposals of UNSC Resolution 242 and called for peace negotiations to begin as soon as possible under "appropriate auspices."[73] On October 22, both Israel and Egypt accepted the cease-fire, and Syria followed a day later. However, up until the cease-fire went into effect fighting continued on both fronts.

With some fits and starts, the cease-fire officially went into effect at 1850 on October 22, the seventeenth day of the war.

The shift in the levels of crisis management that occurred during the first seventeen days of the war was remarkable. The conflict began with only tacit Soviet acceptance and virtually no tactical support for its clients. The cease-fire that ended the fighting, at least temporarily, was authored entirely by the superpowers; their clients were forced to live with the terms. This development made the situation at once more stable and more dangerous. By elevating the crisis to the superpowers level, the more refined crisis-management practices developed over 30 years and formalized by détente made resolution more likely. It was inconceivable, at that point, that the two client states would sit down together and develop between them an acceptable cease-fire. At the same time, the superpowers were now placed in the position of having to guarantee directly the cease-fire and by implication, the behavior of their historically unruly clients. Developments that might occur on the battlefield, even after the cease-fire was implemented, could no longer be artificially segregated into a level of crisis removed from détente. This was to have a profound effect in the course of the subsequent 72 hours.

Tuesday-Wednesday, October 23-24

The cease-fire broke down barely 12 hours after its implementation. Although the responsibility for violating the cease-fire was attributed by each side to the other, it was likely that Israel initiated the violation.[74] Certainly, it was Israel that improved its tactical position as a result of the breakdown in the cease-fire.

Concerned with the fate of the Israeli prisoners held by Egypt and wanting to improve its postwar bargaining position, Israel's strategy became to use the 20,000 Egyptian soldiers and 200 tanks of the Third Army on the east bank of the Sinai as hostages until the Israeli prisoners were returned. To do this, the Suez-Cairo road had to be cut.

The October 22 cease-fire line left the Third Army's LOC still secure because the Suez-Cairo road was in Egyptian hands. Therefore, despite the protestations of both superpowers, Israel continued its drive towards Suez. By the morning of October 24, the Israeli division had pushed through Suez and had completely isolated the Third Army by cutting its last remaining LOC. This cut off the Third Army not only from vitally needed ammunition and replacements, but also from food and water.

Soviet public reaction to the cease-fire violations was apoplectic. In only the second official Soviet statement since the war began, Moscow accused Israel of

> treacherously attacking the positions of the Egyptian forces as well as peacful population points of Egypt. These actions . . . of Israel are an arrogant trampling of a decision of the Security Council and a challenge to the people of the whole world. The Soviet government warns the government of Israel of the most serious consequences which continuation of its aggressive activities . . . will bring.[75]

Unable or unwilling to stop the Israeli drive on Suez, Sadat, on October 24, appealed to the United States and the Soviet Union to honor their commitments to guarantee the cease-fire and intervene to restore the tactical positions that both sides had occupied as of October 22. By accepting the cease-fire, Sadat had turned his local crisis into an issue for direct superpower resolution; he now demanded that the superpowers effect that resolution.

Sadat sent his request to both Moscow and Washington simultaneously. He asked that either the superpowers intervene and push Israel back or "let me push them back myself and please don't regard it as a violation of the cease-fire."[76] Nixon responded to Sadat's request directly:

> Should the two great nuclear powers be called upon to provide forces, it would introduce an extremely dangerous potential for direct great-power rivalry in the area.[77]

The Soviet Union, however, did not share Nixon's reluctance to introduce troops into the conflict. Shortly after Nixon turned down Sadat's request, the United States received another message from Brezhnev that "represented the most serious threat to U.S.-Soviet relations since the Cuban missile crisis."[78] The Soviet Union demanded that the United States impose its will on its client and effect a cease-fire. In the absence of such action, the Soviet Union was prepared to introduce its forces in conjunction with those of the United States or, if necessary, unilaterally.

Nixon and Kissinger urged restraint on the part of Israel but to no avail; Israel maintained its hold on Suez and the LOCs of the Third Army. *Red Star* picked up the earlier Soviet theme and announced again that "most serious consequences would result," if Israel continued to ignore the cease-fire.[79] Moscow put teeth in this threat by altering four more airborne divisions, bringing the total of ready airborne units to seven, the complete Soviet order of battle for airborne divisions. This gave Moscow an immediately available force of some forty to fifty thousand men. Moreover, the Soviet Union curtailed its airlift in order to have ready the necessary airlift assets to transport these units.

At the same time, the Soviet Navy in the Mediterranean stepped up its shadowing of the U.S. Sixth Fleet. By October 24, Soviet naval strength reached 90–100 vessels, over half of which were major combatants.[80]

The United States received yet further indications that a Soviet intervention was increasingly likely. A vigorous exchange of messages between Sadat and Assad, each trying to ascertain whether the other had invited the Soviet Union to intervene, included the Sadat message:

> The Soviet Union has told me that they have sent seventy observers. I understand your position, and accept that you may think it necessary to request Soviet troops if you think the situation calls for them.[81]

This cable was generated by Assad's fears that, if Iraq pulled its forces out of the line, as it was threatening to do if Assad accepted the new cease-fire, a tactical gap would develop, exposing Damascus to an Israeli attack. Heikel speculates that this exchange of messages was intercepted and decoded by the United States, leading to the conclusion that Soviet troops were on their way.

Such a situation was wholly unacceptable to the United States. First of all, it would endanger the American client, Israel, in the conflict. The use of Soviet airborne divisions would surely be followed by more substantial troop commitments, and the Soviet Union would then be in a position to force Israeli units back to the October 22 lines or even farther. This would, in turn, force the United States to commit its forces as well, and the conflict that both superpowers had worked so hard to avoid in Central Europe would develop in the Middle East.

Second, the United States opposed a Soviet intervention because Washington realized that, once in the Middle East, it was unlikely that Soviet troops would ever leave. The lessons of Eastern Europe and the implementation of the Brezhnev Doctrine in Czechoslovakia in 1968 gave ample proof of the willingness of the Soviet Union to use its armed forces to impose its political will. Moscow had been prevented from doing this in the Middle East in the past because of its power projection deficiencies. With a significant military presence in the region, even one introduced ostensibly for peace-keeping purposes, these power projection problems would be largely overcome. The independence of regional leaders, such as Sadat would quickly evaporate.

Third, a Soviet intervention would place Moscow in a position of being able to dictate the terms of the peace to all regional belligerents. To the United States, which had carefully worked its way into a position to exploit politically the military situation, this was unacceptable.[82]

On the basis of the evidence that an intervention was likely and the realization of the steep political price the United States would have to pay for a Soviet presence in the region, Nixon decided that the prevention of a Soviet intervention was worth risking détente and a confrontation with the Soviet Union.

After weighing the evidence, Nixon responded to Brezhnev's frank note of October 24, by saying:

> We must view your suggestion of unilateral action as a matter of the gravest concern involving incalculable consequences. . . . You must know . . . that we could in no event accept unilateral action. . . . Such action would produce incalculable consequences which would be in the interests of neither of our countries and which would end all we have striven so hard to achieve.[83]

To reinforce the seriousness of this message and to place U.S. forces in a position to be able to respond to unforeseen contingencies more quickly, the National Security Council agreed, at 11:30 PM on October 24, to place all U.S. conventional and nuclear forces on increased alert.[84]

This alert, although at only a rather modest level of DefCon Three, reflected the rapid shift of the Middle East conflict from client tanks banging away at each other in the desert with no inherent implications for the international system, to a major confrontation between the nuclear superpowers with the potential to escalate to direct conflict. At this point, with U.S. units on alert, 150 naval combatants shadowing each other in the eastern Mediterranean, nuclear weapons enroute to the Middle East, and seven Soviet airborne divisions prepared to move, the situation by its very nature was dangerous. In such an environment, the political intentions of Moscow and Washington were of marginal importance; war could begin at the hands of a nervous pilot or naval gunner.

Thursday-Saturday, October 25-27

With edgy forces and leaders staring at each other over the sights of locked and loaded weapons, the Soviet Union made one more escalatory gesture. On the morning of October 25, a Soviet freighter declared a transit of the Dardenelles enroute to Cairo. U.S. radiological monitors detected that it was carrying nuclear weapons, and speculation grew in the United States that these weapons were warheads for the Soviet SCUD launchers in Egypt.[85] This was seen as yet another indication of a Soviet intention to intervene.

After that gesture, however, the crisis began to abate. On the battlefield, Israel had completed its investment of the Third Army and required no further combat in order to achieve its objectives. All parties to the conflict had accepted a second cease-fire effective the previous evening, and there were no further reports of significant violations.

Thus, the justification for a Soviet intervention disappeared. With the acceptance of the cease-fire and apparent Israeli intentions to abide by it this time, Sadat began to have second thoughts about the wisdom of inviting the Soviet Union to intervene. In his response to Nixon's letter of the previous day, Sadat agreed that a superpower intervention was probably not a good idea and that he would submit a request to the United Nations for a peacekeeping force.[86] The measure that was introduced to the Security Council on October 25 to establish the United Nations Emergency Force to monitor the cease-fire contained a specific proviso that the composition of the force would not include troops from the permanent members of the Security Council.[87]

The Soviet Union, surprised and alarmed by the rapid escalation in superpower tensions, struck a conciliatory tone. On Friday, October 25, Brezhnev said, "The Soviet Union is ready to cooperate with all interested countries in the cause of normalizing the situation in the Near East."[88] Moreover, the Soviet Union voted for the UNEF mandate in the Security Council, even with the prohibition on troop contributions from the permanent members.

In an effort to extract political points while downplaying the extent of the confrontation, the Soviet press took a ridiculing posture on the U.S. alert. *TASS* said that the statements by U.S. officials that the alert was generated by actions by the Soviet Union were without foundation. The statement went on to say:

> TASS is empowered to declare that such explanations are absurd. . . . This step of the United States . . . was clearly undertaken in an attempt to intimidate the Soviet Union.[89]

Yuri Zhukov, commenting for *Pravda*, continued in the same vein:

> In justification [for the alert] a really absurd allegation was made asserting that the Soviet Union, imagine, the Soviet Union, had taken some kind of action which allegedly gave cause for alarm. . . . What is one to say of the present attempt to intimidate the Soviet Union? It was a time wasting, ungratifying little scheme, one for the scrap heap as they say.[90]

By protesting loudly and strongly that the alert was based on no Soviet provocation whatever, Moscow effectively stepped away from its threat of intervention and its confrontation with the United States.

The SCUD warhead incident also ended quietly, with the nuclear materials returning through the Dardenelles several days after their initial deployment.[91]

The United States, for its part, also tried to move back from the brink. Kissinger stressed that the U.S. alert was only a precautionary measure. Defense Secretary Schlesinger added that, "we were far away from a military confrontation. We were taking actions to preclude a military confrontation."[92] To reinforce this point, the U.S. alert was cancelled on October 27, except for a few units that were involved in training missions.

Aftermath

The crises at both levels thus ended almost as quickly as they had begun. To be sure, there were still many political difficulties that had to be faced in the future, including yet another frank exchange with the Soviet Union over the relief of the still encircled Third Army.[93] But the escalation from the client crisis to a superpower confrontation was effectively over by October 27.

The world, however, was by no means the same. Egypt, despite its net tactical defeat on the battlefield, scored high political points; the bilateral negotiations that were initiated at KM 101 in the Egyptian desert began a process that would result, some six and a half years later, in the signing of a peace treaty with Israel and an Israeli agreement to return the Sinai to Egypt peacefully. War is, after all, an extension of politics by other means, and Sadat's political successes that grew out of the October War were the ultimate judgment of the Egyptian decision in launching the war in the first place.

Egypt's patron, on the other hand, experienced mixed results. Certainly, the disarray in NATO and the oil embargo against the West were substantial, if ephemeral, political gains for the Soviet Union.

But there were also substantial negative implications of the October War for the Soviet Union, implications that bear directly on the understanding of patron-client state relationships in general. The Soviet Union lost control of its client in the war and had to resort to the crudest and most dangerous crisis-management techniques in order to ensure that its client did not suffer a major defeat on the battlefield. It is doubtful that the Soviet Union ever intended to commit its military forces to an intervention in the Middle East. Although Soviet forces are trained to fight in desert environments, the least suitable forces for such operations are the airborne divisions.[94] Like airborne units in all armies, the Soviet airborne divisions are plagued by a lack of organic firepower simply because it is not yet possible to drop a tank or howitzer from an airplane and have it emerge in combat-ready condition. Without overwhelming air superiority and conventional divisions in close proximity, the airborne divisions run the risk of becoming isolated and destroyed. By alerting only the airborne divisions, the Soviet Union would have been preparing the units that were most vulnerable to Israeli armor. Had the Soviet action been accompanied by an alert and movement of armor or mechanized infantry divisions to ports of embarkation, the significance of the Soviet alert would have been far more substantial.

A far more plausible explanation for the Soviet alert was that it was a device to intimidate the United States and force it to put pressure on Israel to accept the cease-fire. As such, it was a successful management of the level-two superpower crisis. Moshe Dayan says:

> The Americans passed on to us evidence that Soviet airborne forces were prepared to intervene directly to save the Egyptian Second and Third Armies. Unless Israel accepted a cease-fire, the United States would not stand in the way of the Soviet Union.[95]

But, although the Soviet management of the patron-level crisis was generally successful and redounded to a net Soviet advantage, it was accomplished at a substantial risk to highly valued Soviet goals in its relationship with the United States.[96] These are risks that the Soviet Union would not have taken, had it not been for the escalation of the client crisis that, in turn, was due to Moscow's inability to manage the crisis at the client level. This inability to manage the crisis at the client level forced the Soviet Union to manage the crisis at the far more dangerous superpower level. Regardless of the political outcome of the crisis, this is a profound lesson to be derived from the October War, one that applies to all patron-client state relationships that involve highly valued patron goals of strategic advantage.

THE DEMISE OF THE RELATIONSHIP

Because Sadat had succeeded politically in his military operation, the days of the Soviet-Egyptian relationship were numbered. The war demonstrated to Sadat that, "The U.S. holds 99% of the cards," in dealing with Egypt's most important security objectives.[97] The demonstrated ability and willingness of the United States first, to force Israel to honor the cease-fire; second, to obtain the First Disengagement Agreement; and third to obtain Israeli cooperation in the Second Disengagement Agreement under which Israel turned over, for the first time, portions of the Sinai to Egypt, convinced Sadat that his future lay with the United States.

It is not necessary here to detail the history of the postwar years of the relationship, for they present only a steady decline in Sadat's attachment to the Soviet Union and an increasing Egyptian willingness to do business with the West and especially the United States. The ultimate demise of the relationship was written in the Egyptian political success of the October War. The reduction in Sadat's threat environment, which occurred as a result of the war, meant that he no longer had to rely on Moscow for weapons. The role of the United States in effecting successive Israeli withdrawals was of far greater value to Sadat in the postwar environment than more Soviet weapons would have been.

The means by which the relationship formally ended stemmed from the shift of the relationship from type III to type VI, a shift due to the reduction of Egypt's threat environment. As discussed in Chapter 2, the client will dominate a type VI relationship and make great demands on the patron in order to honor the patron's goals of strategic advantage. When, in 1975 and 1976, the Soviet Union balked at Sadat's extraordinarily demanding terms, particularly with respect to repayment of Egypt's debts, Sadat abrogated the Treaty of Friendship and Cooperation and, three months later, ejected the Soviet Union from its naval facilities in Egypt. The Soviet-Egyptian patron-client state relationship, begun 21 years earlier, was thus irrevocably ended.

CONCLUSION

The October War displayed, in graphic fashion, the complexity of patron-client crisis management and the two levels at which such crises tend to operate. Both superpowers resolved, throughout the crisis, to keep the war from spilling over into the patron arena. Yet, in the end, because neither Washington nor Moscow could manage the client crisis, it escalated to a superpower confrontation which despite the skepticism that accompanied the various superpower military actions, presented a substantial danger to the international system. Indeed, only by managing the client crisis at the superpower level was Moscow able to effect a resolution of the crisis on terms favorable to its client. Yet, it is this sort of patron confrontation that patron-client state relationships are designed to avoid. The separation of the levels of patron-client crises is thus demonstrated to be tenuous at best and subject to considerable blurring and spillover in the course of a crisis situation.

The lessons derived from the crisis experience of the October War show the meaning and dangers of patron-client state relationships in a nuclear age. The loss of control over a highly valued client with its own goals to pursue has the capacity to involve the patron in direct conflict at the patron level. When the natural instabilities of a crisis are added to the inherent conflicts in patron-client state relationships, the effect can be devastating to the international system.

NOTES

1. The October War has, of course, a variety of names, including the Yom Kippur War, the War of Atonement, and the Ramadan War. All of these names carry some connotation, and therefore, the war will be simply called the October War throughout this discussion.

2. Surprise is one of the three elements that define a crisis, as discussed in Chapter 3. The others are a perception of pressure to resolve the crisis in short order, and the perception that highly valued objectives are at stake.

3. Mohammed Heikel, *The Sphinx and the Commissar* (New York: Harper & Row, 1978), p. 253.

4. This point should not be overblown; the Soviet Union, by its very nature, would never permit the sort of governmental paralysis that accompanied the seizure of the U.S. hostages by Iran.

5. Soviet satellites had no "real time" capability. That is, they could not beam back the images they were taking. Rather, they had to depend upon deorbited "buckets" to transport the film from the satellite to Moscow.

6. Anwar as-Sadat, *In Search of Identity* (New York: Harper & Row, 1977), p. 247. Sadat says that "when we asked about the delay, we were told that the ship had gone astray!"

7. Alvin Z. Rubinstein, *Red Star on the Nile* (Princeton: Princeton University Press, 1977), pp. 263–64. Assad would later deny such coordination, but Rubinstein's arguments are persuasive.

8. Henry Kissinger, *Years of Upheaval.* (Boston: Little, Brown, 1982), pp. 469–70.

9. Just before the June 1967 war, Nasser announced that "our basic objective is to destroy Israel." Theodore Draper, *Israel and World Politics* (New York: Viking Press, 1967), pp. 180–81.

10. Sadat, *In Search of Identity.*p. 255.

11. Kissinger, *Years of Upheaval*, pp. 481–82. Kissinger summarizes the political efficacy of Sadat's limited military objectives and cites the initial Sadat communique on October 7 as evidence.

12. Anwar as-Sadat, *In Search of Identity*, p. 235. It was this factor that Sadat says caused him to call off his planned attack in October 1972. When apprised of the fact that there was a "sand bank gap," Sadat ordered immediate corrective action. By October 1973, the Egyptian banks had been built up to levels higher than their Israeli counterparts.

13. The time of the beginning of the attack was the subject of intense debate between Syria and Egypt. Syria wanted the attack to occur early in the morning in order to have the rising sun in the eyes of the Israeli defenders. Using the same logic, Egypt wanted the attack to begin late in the day. The issue was resolved on October 3, through the personal intervention of Syrian president Assad (Mohammad Heikel, *The Road to Ramadan* [New York: Quandrangle/The New York Times Book Company, 1975], p. 31). This discussion will not dwell on the Soviet-Syrian connection, although that relationship is, in itself, another useful case study of patron-client state relationships. Syria will be considered only inasmuch as it impacts on the Soviet-Egyptian relationship.

14. Israeli and American intelligence had been aware of Arab war preparations, although they both misjudged the timing and intensity of the attack. Only on the actual day of the conflict did Israel fully understand that an attack was imminent. Moshe Dayan and Golda Meir claimed that Israel knew of the attack but elected not to preempt because of the adverse impact this would have on world opinion. It is more likely, given the historic disregard for world opinion that has marked Israeli policy, that Israel realized that it could do little to preempt the Arab attack, given the late date.

15. The Syrian Army, like Egypt, is organized along Soviet lines. A Syrian division is, therefore, about half the size of a U.S. or Israeli division.

16. Kenneth S. Brower, "The Yom Kippur War," in *Military Review*, March 1974, p. 26.

17. Egypt used high-pressure water hoses, purchased in the FRG, to wash the loose sand away from the east bank in order to provide level platforms for their heavy bridges. This tactic was wholly an Egyptian invention.

18. Heikel, *The Road to Ramadan*. p. 44.

19. Ibid. p. 211.

20. Chaim Herzog, *The War of Atonement* (Boston: Little, Brown, 1975), pp. 156, 164, 168. Herzog's reporting of the battle along the Suez front is substantially at odds with both Heikel and Sadat. Therefore, his presentation of Israeli losses has particular credibility.

21. Sadat, *In Search of Identity*. pp. 252–53.

22. Heikel, *The Road to Ramadan*. p. 209.

23. Brower, "The Yom Kippur War," p. 15.

24. Sadat, *In Search of Identity*. p. 253.

25. To some extent, discussions of the cease-fire proposal raise artificial expectations. It is inconceivable that Israel would have accepted any cease-fire that left it at a tactical disadvantage.

26. Sadat, *In Search of Identity*. p. 253. With some asperity, Sadat reports that this ammunition was that which the Soviet Union failed to deliver during the war of attrition four years earlier. The Soviets used this argument with Kissinger to downplay the significance of the airlift (Kissinger, *Years of Upheaval*, pp. 549–50).

27. Jon D. Glassman, *Arms for the Arabs* (Baltimore: The Johns Hopkins University Press, 1975), p. 143.

28. Lester A. Sobel, ed., *Israel and the Arabs: The October 1973 War* (New York: Facts on File, 1974), p. 94.

29. Brower, "The Yom Kippur War," p. 27.

30. Herzog, *The War of Atonement*. p. 145. Herzog reports that total Israeli losses for the Golan campaign were 250 tanks and 772 personnel. He also provides Syrian loss figures of a slightly smaller magnitude.

31. *Department of State Bulletin*, October 29, 1973, p. 529; Kissinger, *Years of Upheaval*. pp. 490–91.

32. Sadat, *In Search of Identity*. p. 254.

33. Heikel, *The Road to Ramadan*. p. 213.

34. Sadat refused to believe in any limitations in the abilities of the superpowers to monitor the battle. He was convinced that both the United States and the Soviet Union knew exactly what was going on before he did.

35. Heikel, *The Road to Ramadan*. p. 214.

36. Ibid., p. 215.

37. *New York Times*, October 10, 1973; Marvin Kalb and Bernard Kalb, "Twenty Days in October," in *The New York Times Magazine*, June 23, 1974.

38. Cited in Rubinstein, *Red Star on the Nile*. p. 266.

39. Heikel, *The Road to Ramadan*. p. 214.

40. Ricard Nixon, *The Memoirs of Richard Nixon* (New York: Grosset & Dunlap, 1978), p. 922.

41. Heikel, *The Road to Ramadan*. p. 218.

42. *Time*. October 22, 1973, p. 51 (reprinting an article from *Pravda*, October 11, 1973).

43. Samuel W. Sax, and Avigdor Levy, "Arab-Israeli Conflict Four: A Preliminary Assessment," *Naval War College Review*, January-February 1974, p. 13.

44. Secretary of Defense James Schlesinger, in his news conference of October 26, 1973, as reported in the *Department of State Bulletin*, November 12, 1973, p. 619.

45. Herzog contends that Egyptian armor was deployed across the Canal on the second day of the war. However, since most of the armor had been attached to the First Army, it is unlikely that this was true.

46. Sadat, *In Search of Identity*. p. 255.

47. Cited in Rubinstein, *Red Star on the Nile*, pp. 269–70; Kissinger, *Years of Upheaval*, pp. 507–08, contains a compelling rationale for moderate language, reinforcing the two-levels concept.

48. Richard Nixon, *The Memoirs of Richard Nixon* (New York: Grosset & Dunlap, 1978), p. 927; Marvin Kalb, and Bernard Kalb, "Twenty Days in October," *The New York Times Magazine*, June 23, 1974, pp. 50–54.

49. Kissinger, *Years of Upheaval,* p. 507. Kissinger outlines several actions eschewed by Moscow in the name of preserving détente.

50, Cited in Glassman, *Arms for the Arabs,* pp. 148–49.

51. Ibid., p. 134.

52. Brower, "The Yom Kippur War," p. 28. Hussein is much maligned in the Arab world for not opening the West Bank front. However, there is probably no more difficult terrain in the world to attack over than the Judean hills; a Jordanian attack in that area would have been useless.

53. Heikel, *The Road to Ramadan,* p. 225.

54. Sadat, *In Search of Identity.* p. 258.

54. Of the NATO allies, only Portugal granted overflight permission to the United States. One ally, Turkey, refused to allow U.S. overflights but permitted such overflights by the Soviet Union.

56. Rubinstein, *Red Star on the Nile,* p. 272.

57. Herzog, *The War of Atonement,* p. 206. He reports that some 246 Egyptian tanks were destroyed in close combat alone on October 14.

58. Heikel, *The Road to Ramadan,* p. 228.

59. Herzog, *The War of Atonement,* pp. 208–09.

60. Cited in Rubinstein, *Red Star on the Nile,* p. 270.

61. Ibid., p. 271.

62. Sadat turned the day-to-day management of the war almost entirely over to his professional soldiers, a remarkable departure from the conduct of most political leaders.

63. Heikel, *The Road to Ramadan,* p. 231.

64. Kalb, and Kalb, "Twenty Days in October," p. 54.

65. Sadat, *In Search of Identity,* p. 259.

66. Nixon, *The Memoirs of Richard Nixon,* p. 928. On October 19, Nixon requested an emergency $2.2 billion to replace Israeli battle losses.

67. This engagement is usually referred to as the Battle of the Chinese Farms. When Israel overran the Sinai in 1967, its troops found experimental agricultural enterprises established by Japan. The soldiers mistook the Japanese letters for Chinese.

68. Sadat, *In Search of Identity,* p. 259.

69. Ibid. Most other sources report that Kosygin returned to Moscow with Sadat's agreement in his pocket. Because of the source, however, more credence needs to be given to Sadat's own version. Sadat's acceptance was not public at this time.

70. Ibid.

71. Nixon, *The Memoirs of Richard Nixon,* p. 931.

72. Heikel, *The Road to Ramadan,* p. 237. Heikel admits to eavesdropping on a conversation between Sadat and Ismail.

73. Rubinstein, *Red Star on the Nile,* p. 274.

74. Kissinger, *Years of Upheaval,* pp. 508–09.

75. Cited in Glassman, *Arms for the Arabs,* p. 158.

76. Sadat, *In Search of Identity,* p. 266.

77. Nixon, *The Memoirs of Richard Nixon,* p. 938.

78. Ibid.

79. Martin J. Slominski, "The Soviet Military Press and the October War," *Military Review,* May 1974, p. 40.

80. *New York Times,* October 26 and October 30, 1973.

81. Heikel, *The Road to Ramadan,* p. 253.

82. Kissinger, *Years of Upheaval,* pp. 583–84.

83. Nixon, *The Memoirs of Richard Nixon,* pp. 939–40.

84. There is some doubt as to whether or not Nixon was actually present when the decision was made. Certainly, he implies that he was. However, other sources report that the president, preoccupied by the mounting Watergate situation and the "Saturday Night Massacre," went to bed, leaving the NSC in the Situation Room to work on the problem (Nixon, *The Memoirs of Richard Nixon,* p. 939; Kalb and Kalb, "Twenty Days in October," p. 61; Rubinstein, *Red Star on the Nile,* pp. 275–76; Kissinger, *Years of Upheaval,* pp. 585–91).

85. Rubinstein, *Red Star on the Nile,* p. 276.

86. Nixon, *The Memoirs of Richard Nixon,* p. 940.

87. Glassman, *Arms for the Arabs,* p. 165.

88. Ibid.

89. Ibid., p. 166.

90. Ibid., p. 167.

91. The entire issue of the nuclear warheads remains murky. Some, such as Senators Symington and Stennis, speculate that no weapons ever did make the Dardenelles transit, and Kissinger confirms that the U.S. had no firm evidence. In any event, the warheads never became a factor, beyond the political symbolism their deployment had.

92. Glassman, *Arms for the Arabs,* p. 164.

93. Nixon, *The Memoirs of Richard Nixon,* p. 942; Kissinger, *Years of Upheaval,* pp. 601–11, provides a lucid account of the Third Army Issue.

94. Graham H. Turbville, "Soviet Desert Operations," *Military Review,* June 1974, pp. 40–41.

95. Glassman, *Arms for the Arabs,* pp. 164–65. Dayan made many mistakes in the October War and often searched for ways to shift the responsibility for the negative aspects of the war elsewhere.

96. William E. Odom. "Whither the Soviet Union," *The Washington Quarterly,* Spring 1981, p. 31. Odom argues that the demise of détente began with the Soviet activities in the October War.

97. Cairo Domestic Service, October 15, 1977.

CHAPTER 7

THE RELATIONSHIP REVISITED

In preceding chapters. we have identified certain prominent features of patron-client state relationships, the goals that give them form, and the mechanisms by which they change. These features were applied to individual cases and then to the 20-year history of the relationship between Egypt and the Soviet Union. Because such relationships involve complex interactions and seemingly divergent analytical threads, it is important, by way of summary, to outline the prominent features, the underlying goal structures, and the implications of the primary mechanisms of change of such relationships.

FOUNDATIONS OF THE RELATIONSHIP

Throughout the above discussion, it is apparent that, at their foundation, patron-client state relationships rest upon two elements that shape their nature and dictate their impact on the international system as a whole.

First, patron-client state relationships are inherently unstable. This stems from the fundamental incompatibility of basic patron and client objectives. The patron seeks to exercise some measure of control over the client's political, economic, or geographic resources. The client, on the other hand, wants to tap into the patron's security resources without surrendering any measure of its autonomy. This means that the patron and client are perpetually at odds with each other.

This incompatibility is reinforced by the dramatically different world view patron and client maintain and relative scope of their political vision. The patron, especially if it is one of the superpowers, looks at the international system from a global perspective. The patron is concerned with a plethora of relationships spanning all the world's political subsystems. The patron usually understands, if only implicitly, the interactions between the different regions and its different bilateral relationships. The patron's primary security calculus is shaped by the realities of nuclear weapons; only through the use of such weapons is the patron's national survival put at risk. As a result of 30 years of experience with managing these nuclear realities, patrons are not routinely confronted with issues of national survival.

The client, by contrast, has a much smaller field of political vision and a much more limited understanding of how the various parts of the international system relate to one another. More importantly, the client is more often confronted with threats to its national survival stemming from a variety of sources. To the client, the threat of global nuclear incineration, so central to the patron's security calculus, is an esoteric danger; it is far more concerned with regional forces on its borders that can destroy it as surely as a nuclear exchange.

Patron-client state relationships are, thus, substantially different from patron-patron interactions; client states are not simply small patrons. They operate from a different frame of reference, with no common appreciation of the global implications of superpower confrontation. Indeed, when the client perceives that its own national existence is in jeopardy, it has little sympathy for patron fears of nuclear conflict.

Because of this first basic feature of patron-client state relationships, they are susceptible to dramatic change over time. The ebb and flow of influence between patron and client make such relationships generally ill defined and subject to constant misinterpretation.

The second fundamental feature of patron-client state relationships is that they are, at their most basic level, mechanisms by which patrons compete with each other in a theoretically low-risk environment. Since direct patron competition, especially when it involves superpowers, carries with it the risk of rapid escalation to nuclear conflict, patrons normally eschew such direct confrontation. Rather, they attack each other by using their clients as tools for furthering their competitive interests and for obtaining unilateral advantage. Such surrogate competition is attractive because it takes place, again in theory, on the periphery of the international system far from the mainstream of intra-patron relations. Patron setbacks or mistakes in these areas do not, in themselves, place at risk central patron objectives or involve issues of national survival.

This second fundamental characteristic of patron-client state relationships forms the basis for the specific objectives the patron seeks. These objectives, in turn, dictate the patron's valuation of the relationship and the price the patron is willing to pay to maintain it. The more substantially the client can contribute to the patron's competitive advantage over its adversaries, the more the patron will be willing to invest in the relationship and the more ready the patron will be to accede to client demands. In this regard, relationships in which the patron perceives the possibility of realizing goals of strategic advantage (types III and VI) will be particularly valued, and the patron will be more susceptible to rapacious client demands. Under such circumstances, the patron may even surrender a substantial measure of its political and military flexibility to the client in order to meet client demands, to ensure that the relationship endures, and to protect the strategic advantage goals the patron perceives.

When goals of strategic advantage dominate the patron's calculus, the client may thus assume a position of great influence in the relationship. This is particularly true in type VI relationships in which the client has a low-threat environment. The client, unpressured by events, will feel more able to identify and pursue specific or economic objectives with its patron in a thoughtful, rational manner.

When the patron has strategic advantage goals and the client faces a high-threat environment (type III), the flow of influence will move back and forth. Relationships of this type are especially unpredictable and unstable, with concomitant dangers of rapid, unexpected escalation in a crisis environment. These are the most dynamic and explosive of the various patron-client state relationship types.

The goals the patron pursues, coupled with the nature of the client's threat environment, determine the relationship type at any given moment in history. This, in turn, determines the potential impact that that a specific relationship can have on the international system as a whole. In general, the more influence the client can exert and the greater the client's access to the patron's security resources is, the greater the danger to the international system a particular patron-client state relationship will pose. This facet is rooted in the basic differences in world view and historical experience between patron and client outlined above. The client is generally incapable of managing the patron's resources in a responsible manner, yet some types of relationships can place considerable control in the client's hands.

The relationship type at any point in time may not long endure; patron-client state relationships are highly dynamic. Change can occur in such relationships because of changes in the patron's goal structures, the client's threat environment, or the international competitive milieu. This propensity for change is reinforced by the basically unstable nature of patron-client state relationships and the fundamentally conflicting objectives both states have.

Thus, the two basic elements of patron-client state relationships, their inherently unstable nature and their role in intra-patron competition, shape the dynamism and importance of such relationships in the international system. These are the common threads that run through all of the examples cited in earlier chapters and are particularly evident in the relationship between the Soviet Union and Egypt.

CRISIS AND PATRON-CLIENT STATE RELATIONSHIPS

Patron-client state relationships and crisis management are not normally thought of as being closely related in most studies on each subject. However, as is evident from the discussion in this study, a rigorous, meaningful examination of patron-client state relationships is impossible without, at the same time, dealing at length with crises. These two normally divergent threads of political analysis are, in fact, inextricably interwoven into a single fabric of political reality. As such, they must be considered together. To be sure, there are crises that do not require the study of patron-client state relationships to be understood. The converse, however, is not true; crises are integral parts of the richness and complexity of patron-client state relationships and must be thoroughly under-stood and appreciated.

The most volatile and pervasive mechanism for change in patron-client state relationships is, in fact, the crisis. Crises cause both patron and client to make rapid and unstudied decisions based upon either inadequate or unrefined information. Coupled with this is the lack of understanding of the fundamentals of patron-client state relationships and the role of crises in their dynamics. This results in hurried, pressured, and sometimes irrational policy decision making. It also results in the high potential for client crises to escalate to patron conflict, a phenomenon that lies at the root of the real dangers to the future of the international system.

Because crises are so integral to patron-client state relationships and because crises present the greatest dangers of escalation into global conflict, analysts and policymakers alike must fully consider the unique nature of crises, and their management aspects, in the patron-client context.

Levels of Crises

As was demonstrated during the October 1973 War, crises in patron-client state relationships proceed at several levels, often simultaneously. First, crises occur at the client level as Third World states pursue their own national priorities and security objectives, which may or may not be compatible with the desires of their patrons. Crises at the client level in themselves stress patron-client state relationships and are the principal mechanisms of relationship change, as was outlined in Chapter 3 and demonstrated repeatedly during the Soviet-Egyptian relationship. Were a crisis to remain solely at the client level, it would remain crucial in shaping the character of the relationship itself but would not pose dangers to the larger international system.

But crises in the patron-client context can also occur at the level of patrons themselves. A client crisis can readily escalate to a patron confrontation in which the client crisis itself may become only a secondary issue. The apparently easy transition from a client-level to a patron-level crisis is dictated by the fundamental role patron-client state relationships play in intra-patron competition. No patron, especially a superpower, can afford to have its client defeated at the hands of a client of an adversary patron. To do so would be to lose ground in intra-patron competition, perhaps in a substantial manner, depending on how valuable the client is.

When patron-client crises flow across the two crisis levels, complex combinations of interactions can occur, as was demonstrated in Chapter 5. In the case of the October War, the initiating crisis was between the two clients. It then involved extensive interactions between the patrons and their respective clients (in terms of resupply and advice) and between patrons and their client's adversaries (in terms of Soviet threats and exchanges of letters between Sadat and Nixon). Finally, the client crisis escalated into a patron confrontation, with all the inherent dangers to the international system such a confrontation involves. This is portrayed schematically in Figure 7.1.

The lessons learned from the complexity of crisis interactions during the October War are applicable to the entire spectrum of patron-client relationships. Adding yet further complications to crises in patron-client relationships are factors that are exogenous to the immediate crisis but factor in strongly in patron and client crisis calculi.[1]

Crisis Management

Figure 7.1 demonstrates the complexity of managing crises in patron-client state relationships. Yet, because crises are such a pervasive element of such relationships, crisis management becomes a crucial skill. Because of their role in intra-patron competition and because of the propensity for crisis escalation, patron-client state relationships present a unique set of crisis-management challenges for patron decision makers and their policy structures.

To be effective, crisis management in patron-client state relationships must be successful on the two crisis levels. First, the patron must deal with the immediate crisis in which the client finds itself. This involves a clear understanding by the patron of the nature of the goals it seeks in the relationship and the extent of the investment it is prepared to make to maintain the relationship. It also involves extensive precrisis planning in order to establish various mechanisms for monitoring the crisis once it begins, for controlling the actions of the client, and for avoiding crisis escalation.

Second, the patron must deal with the bilateral crisis that may develop between patrons as a result of escalation of the client crisis. To be sure, in some client crises, the escalation potential may be relatively minor; there may be no other patron directly involved.

Figure 7.1

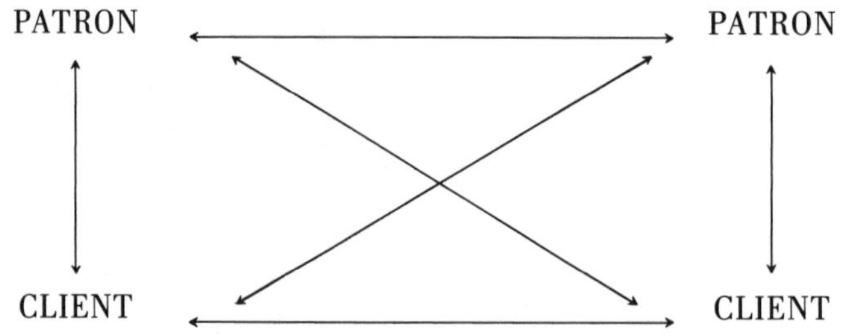

However, such situations are generally the exception. Most client crises involve, in one way or another, more than one patron. The involvement of several patrons with a crisis transfers the crisis to the patron level. Moreover, many of the actions a patron may take to resolve the crisis in the client's favor may, in fact, exacerbate the crisis at the patron level. Patrons involved in an escalating client crisis must, therefore, keep their security priorities clearly in mind.

The two levels of crisis management thus do not separate themselves neatly into discrete policy problems; the complex interaction possibilities presented above mean that there is considerable movement back and forth between the different levels. The bilateral patron crisis, because of its roots in the generating client crisis, is not readily amenable to established patron crisis-management techniques. The clients, through actions of their own, retain the capability to upset the most carefully orchestrated patron crisis-management efforts. This was clearly the case in the October War when Israeli violations of the October 22 cease-fire precipitated the superpower confrontation that, in turn, was defused only when the unruly clients agreed to behave themselves and adhere to the cease-fire.

The extent of the linkage between the levels of a patron-client state relationship crisis is dependent upon the nature of the relationships involved. In cases where type III relationships are at issue, such as was the case in the October War, the linkage between the crisis levels and the complexity of crisis management are quite high. We would expect the same in an India-Pakistan conflict because both clients are involved in type II or type III relationships with their respective patrons. In cases where the relationships are of a lower order, the risks of spillover from one crisis level to the other would be considerably smaller. A war between the Congo and Zaire, which would involve type II clients, would not run a great risk of superpower confrontation.

To be effective, patron-client crises must be managed at both levels simultaneously. Policymakers who anticipate managing a client crisis only at that level will be rudely shocked to discover that other patrons may become involved quite rapidly. The outbreak of a client crisis should, therefore, elicit an immediate effort to establish and maintain crisis-management efforts at both levels.

Crisis Manipulation

Compounding the already complex crisis-management problem is the reality of crisis manipulation discussed at length in Chapters 3 and 4. Because crises are such powerful tools in shaping the nature of patron-client state relationships, either partner may attempt to exaggerate or even fabricate a crisis in order to move the basic character of the relationship in a direction it perceives as more favorable to its interests. This means that the states involved in a crisis may not all be interested in reducing tensions and managing the crisis. One or more states may, in fact, try to thwart the crisis-management efforts of others in order to keep the crisis going in the direction they desire. Soviet efforts in May 1967 to undercut U.S. attempts to calm the Middle East were reflective of this aspect of crisis manipulation. Syngman Rhee's behavior in trying to undermine the Korean armistice is another example.[2]

Crises, their management and manipulation, are thus crucial to understanding the nature of patron-client state relationships. Their role and importance in such relationships cannot be ignored if meaningful analysis is sought.

PATRON-CLIENT STATE RELATIONSHIPS IN THE FUTURE

As has been stressed throughout this study, patron-client state relationships are fundamentally tools for patron competition. Because of this, the role of such relationships in the future will be a function of the competitive nature of the international sytem in the decades to come.

There is no evidence to suggest that competition between the superpowers or between other patrons will abate in the foreseeable future. Indeed, the failure of détente to reduce superpower competition during the late 1960s and 1970s augurs poorly for closer cooperation in the future. The United States, under the leadership of Nixon and Kissinger, felt that by involving the Soviet Union in a web of interdependent relationships, Soviet adventurism and expansionism could be controlled. Under the Kissinger approach, the

Soviet Union would be unwilling to risk its ties with the West by aggression in the Third World and would be forced to control its clients for the same reason.[3] To support this approach, the United States provided the Soviet Union with a host of scientific exchanges, technology transfers, economic incentives, and political demonstrations over the course of the decade of détente.[4]

The Soviet invasion of Afghanistan, however, largely discredited the cooperative assumptions of détente. The Carter administration determined that the Soviet Union was not moderating its behavior in the international arena nor was its drive to compete with the West diminishing. The subsequent election of Reagan administration confirmed the competitive outlook of the United States.

With the collapse of the cooperative underpinnings of détente, the sublimated superpower rivalries, never far below the surface, became the primary feature of the international system again, checked only by the mutual fear of a general war. This situation will, in all likelihood, continue to dominate until a major internal realignment occurs within one of the primary power blocs or a nuclear war creates a new international order.

We should anticipate, therefore, an international system marked by what Odom has described as "competitive engagement" in which the superpowers, and their associated patron allies, will increasingly compete for power and influence in a variety of channels.[5]

In this sort of environment, patron-client state relationships will assume mounting significance, particularly in Latin America, Africa, Southwest Asia, and the Middle East. The superpowers will continue to attempt to accrue clients through which they can gain some form of advantage over their competitiors, and each superpower will continue to attempt to undermine the key relationships of its opponents in an effort to place them at a disadvantage.

Competiton in the future will not, of course, be limited to the superpowers. The client states will continue their own struggles for regional dominance and for national survival. This competition will be further fueled by the influx of highly sophisticated weapons from the superpowers and other patrons, weapons that will help extend the capabilities of smaller states to engage in international mischief. Once again, the trends are in the direction of greater instability in the Third World which, in turn, will present greater opportunities for patron states to exploit unstable situations and to extend their competitive activities.

All of this means that patron-client state relationships will become an even more important element of the international system in the future. Understanding the key ingredients of these relationships and the mechanisms by which they change, is therefore critical to both meaningful political analysis and effective policy formulation in the decades ahead.

CONCLUSION

This study has attempted to focus on the political realities of patron-client state relationships. At their most basic level, these relationships provide mechanisms by which patron states, and especially the superpowers, can compete with each other while avoiding the dangers of direct confrontation. Patron-client state relationships cannot be meaningfully studied without reference to the more general systemic competition they support.

The flow of influence in the relationship is dependent upon the nature of the objectives sought by both patron and client. In those relationships in which the patron seeks goals of only marginal utility in its competition with other patrons, the flow of influence will normally favor the patron, particularly in type I relationships in which the client faces a high-threat environment. Because patron states are generally more proficient at crisis management and more aware of the implications of confrontation for the entire international system, relationships of this type are normally benign in terms of their impact of international stabililty and nuclear security. At the other extreme, when the client has a low-threat environment and the patron seeks very highly valued goals of strategic advantage, the client can be expected to maintain a large measure of influence over the patron and can exert substantial control over the patron's diplomatic and military capabilities. These are relationships that pose perhaps the greatest threat to international security in the nuclear age and will continue to do so in the future, particularly as the superpower competitive environment intensifies.

To the extent that patron-client state relationships help reduce the probability of direct confrontation between the superpowers by providing an outlet for their competitive drives, such relationships are a stabilizing force in the international system. However, when these relationships result in the loss of patron control and the escalation of client crises into patron confrontation, the implications for the international system can be profoundly destabilizing.

Whatever normative judgments one might reach about patron-client state relationships, the reality of their role in the present community of nations cannot be ignored, particularly as we face an increasingly competitive environment in the future. These relationships must be understood and policies shaped accordingly in order to protect the international system from their potentially destructive aspects.

To the extent that this study contributes to understanding patron-client state relationships, their underlying dimensions, and their crucial linkages to crises, this has been a useful enterprise. Much more, however, remains to be done.

NOTES

1. Environment factors affecting decision making are outlined in Harold L. Sprout and Margaret Sprout, "Environmental Factors in the Study of International Politics," in *International Politics and Foreign Policy*, ed., James N. Rosenau, (New York: The Free Press, 1969), pp. 44–56.

2. Chang Jin Park, "The Influence of Small Powers on the Superpowers: United States-South Korea Relations as a Case Study, 1950–53," in *World Politics*, October 1975, pp. 108–12.

3. Robert G. Kaiser, "U.S.-Soviet Relations: Goodbye to Détente," in *Foreign Affairs*, America and the World, 1980, p. 501

4. William E. Odom, "Whither the Soviet Union," in *The Washington Quarterly*, Spring 1981, pp. 43–44.

5. Odom, "Whither the Soviet Union," p.46. This was to be the theme of the second Brzezinski NSC.

BIBLIOGRAPHY

BOOKS AND ARTICLES

The Air Defense Enhancement Package for Saudi Arabia. Washington: The National Security Council, 1982.

Ajami, Fouad. "On Nasser and his Legacy." *Journal of Peace Research* 11, (1974): 41–50.

Art, Robert J. "To What Ends Military Power." *International Security* 4 (Spring 1980).

Askori, Hassein, and Vittorio Corbo. "Economic Implications of Military Expenditures." *Journal of Peace Research* 11 (1974): 341–44.

Aspaturin, Vernon V. "Soviet Global Power and the Correlation of Forces." *Problems of Communism* (May-June 1980): 1–18.

Bar-Simon-Tov, Yaacov. *The Isreali-Egyptian War of Attrition, 1969–1970.* New York: Columbia University Press, 1980.

Barclay, C.N. "Lessons from the October War—Learning the Hard Way." *Army* 24 (March 1974): 25–32.

Battle, Lucius D. "The Arabs: Why Now?" *The New York Times Magazine*, October 21, 1973.

Bell, Coral, *The Conventions of Crisis.* (New York: Oxford University Press, 1971).

Blechman, Barry M., and Stephen S. Kaplan. *Force Without War.* Washington: The Brookings Institution, 1978.

Brady, Linda P., and Ilan Peleg. "Carter's Policy on the Supply of Conventional Weapons: Cultural Origins and Diplomatic Consequences." *Crossroads*, 5 (Winter 1980), pp. 41–68.

Brecher, Michael. "The Subordinate State System of Southern Asia." In *International Politics and Foreign Policy*, edited by James N. Rosenau. New York: The Free Press, 1969.

Bretton, Henry L. *Patron-Client Relations: Middle Africa and the Powers.* New York: General Learning Press, 1971.

Brezhnev, Leonid. In *Vital Speeches* 39 (July 15, 1973): 578–80.

Brower, Kenneth. "The Yom Kippur War." *Military Review* 54 (March 1974): 25–33.

Buckley, William F., Jr. "Human Rights and Foreign Policy: A Proposal." *Foreign Affairs* 58 (Spring 1980): 775–96.

Campbell, John C. "The Communist Powers and the Middle East."*Problems of Communism* 21 (September-October 1972): 40–53.

Chouci, Nazli. "International Nonalignment." In *International Systems*, edited by Michael B. Haas. New York: Chandler, 1974.
Chubin, Shahram "U.S. Security Interests in the Persian Gulf in the 1980s." *Daedalus* 109 (Fall 1980): 31-66.
Communist Aid to Less Developed Countries of the Free World. Washington: The Central Intelligence Agency, 1978.
Congressional Presentation. Washington: The Defense Security Assistance Agency, 1980 and 1981.
Cooley John K. "The Shifting Sands of Arab Communism." *Problems of Communism* 24 (March-April 1975): 22–42.
Corcoran, Edward A. "Soviet Muslim Policy: Domestic and Foreign Policy Linkages." Paper presented at the U.S. Army War College Strategic Studies Institute Military Policy Symposium, September 1979 Carlyle Barracks, Penn.
Coddry, Charles W. "The Yom Kippur War, 1973—Lessons New and Old." *National Defense* 58 (May-June 1974): 505–08.
Cottrell, Alvin J., and R. M. Burrell. "The Soviet Navy and the Indian Ocean." *Strategic Review* 2 (Fall 1974): 25–38.
Cromwell, William C. "Europe and the 'Structure of Peace.' " *Orbis* 22 (Sring 1978): 11–36.
Crump, Roger L. "The October War: A Postwar Assessment." *Military Review* 54 (August 1974): 12-26.
Daniel, Donald C. "Sino-Soviet Relations in Naval Perspective." *Orbis* 24 (Winter 1981): 787–803.
David, Steven. "Realignment in the Horn: The Soviet Advantage." *International Security* 4 (Fall 1979): 69–90.
Deutsch, Karl, and J. David Singer. "Multipolar Power Systems and International Stability." In *International Politics and Foreign Policy*, edited by James N. Rosenau. New York: The Free Press, 1969.
DeVore, Ronald M. "The Arab-Israeli Arms Race and the Superpowers." *Current History* 66 (February 1974): 70–73.
Dowdy, W. L. "The Politics of Oil in the War of Yom Kippur." *U.S. Naval Institute Proceedings* 100 (July 1974): 23–29.
Draper, Theodore. *Israel and World Politics.* New York: Viking Press, 1967.
Dur, Philip A. "U.S. Sixth Fleet: Search for Consensus." *U.S. Naval Institute Proceedings* 100 (June 1974): 18–23.
Eilts, Hermann F. "Security Considerations in the Persian Gulf." *International Security* 5 (Fall 1980): 79–113.
Eran, Oded. "Soviet Middle East Policy, 1967–1973." In *From June to October*, edited by Itamar Rabinovich, and Haim Shaked. New Brunswick, N.J.: Transaction Books, 1978.
Farer, Tom J. "Searching for Defeat." *Foreign Policy*, 40 (Fall 1980), pp. 155–174.

Fehrenback, T. R. *This Kind of War*. New York: Macmillan 1963.

Fitzgerald, Frances. *Fire in the Lake*. Boston: Little, Brown, 1972.

Foreign Military Sales and Military Assistance Facts. Washington: The Defense Security Assistance Agency, 1978.

Freedman, Robert O. *Soviet Policy Toward the Middle East Since 1970*. New York: Praeger, 1975.

Friedgut, Theodore H. "The Middle East in Soviet Global Strategy." *The Jerusalem Journal of International Relations* 5 (1980): 66–93.

Funkhoser, John T. "Soviet Carrier Strategy." *U.S. Naval Institute Proceedings* 99 (December 1973): 27–37.

Geiger, Theodore, and Neil J. McMullen. "Soviet Options in the Persian Gulf and U.S. Responses." *New International Relations* 5 (July 1980): 7–17.

George, Alexander, David K. Hall, and William R. Simons. *The Limits of Coercive Diplomacy*. Boston: Little, Brown 1971.

George, Alexander, and Richard Smoke. *Deterrence in American Foreign Policy: Theory and Practice*. New York: Columbia University Press, 1974.

Glassman, John D. *Arms for the Arabs*. Baltimore Johns Hopkins University Press, 1975.

Golan, Galia. "Syria and the Soviet Union since the Yom Kippur War." *Orbis* 21 (Winter 1978): 777–99.

Goldman, Marshall I. "Is There a Russian Energy Crisis." *The Atlantic* 246 (September 1980): 55–64.

Goure, Leon, and Micheal J. Deane. "The Soviet Strategic View." *Strategic Review* 8 (Spring 1980): 101–115.

Griffith, William E. "The Implications of Afghanistan." *Survival* (July-August 1980): 146–51.

Hagan, Kenneth J., and Jacob W. Kipp. "U.S. and U.S.S.R. Naval Strategy." *U.S. Naval Institute Proceedings* 99 (November 1973): 38–44.

Hammond, Paul Y., David J. Louscher, and Michael D. Salomon. "Controlling U.S. Arms Transfers." *Orbis* 23 (Summer 1979): 317–52.

Handel, Michael I. "Surprise and Change in International Politics." *International Security* 4 (Spring 1980): 57–85.

Hartman, Frederick H. *The Relations of Nations*. New York: Macmillan 1978.

Heikel, Mohammed. *The Cairo Documents*. New York: Doubleday and Company, 1973).

________. *The Road to Ramadan*. New York: Quadrangle, 1978.

________. *The Sphinx and the Commisar*. New York: Harper & Row, 1978.

Hermann, Charles F. "International Crisis as a Situational Variable." In *International Politics and Foreign Policy*, edited by James N. Rosenau, New York: The Free Press, 1969.

Herzog, Chaim. *The War of Atonement*. Boston: Little, Brown, 1975.

Hirschfeld, Yair P. "Moscow and Khomeini: Soviet-Iranian Relations in Historical Perspective." *Orbis* 24 (Summer 1980): 219–40.

Hoffman, Stanley "Requiem." *Foreign Policy* 42 (Spring 1981), pp. 3–42.

Horowitz, Irving L. "The Mid-east, Peace or No More War?" *Current*, no. 160 (March 1974), pp. 51–55.

Hottinger, Arnold. "Soviet Influence in the Middle East." *Problems of Communism*, 24 (March-April 1975), p. 72.

Howard, Henry N. "The Soviet Union in Lebanon, Syria, and Jordan." In *The Soviet Union in the Middle East* edited by Ivo Lederer and Wayne S. Vucinich. Stanford: The Hoover Institution, 1974.

Howard, Michael. "Return to the Cold War." *Foreign Affairs* 58 (America and the World, 1980): 459–73.

Hudson, George E. "Soviet Naval Doctrine and Soviet Politics." *World Politics* 29 (October 1976): 90–113.

Hughes, Thomas L. "The Crack-up: The Price of Collective Irresponsibility." *Foreign Policy*, 40 (Fall 1980), pp. 33–60.

Husband, William B. "Soviet Perceptions of the U.S. 'Position of Strength' Diplomacy in the 1970s." *World Politics* 31 (July 1979): 495–17.

Irani, Robert Ghobad. "Changes in Soviet Policy Toward Iran." Paper presented at the U.S. Army War College Strategic Studies Institute Military Policy Symposium, September 1979, Carlyle Barracks, Penn.

Issues Concerning the Proposed Sale of Airborne Warning and Control E-3 Aircraft to Iran. Washington: The General Accounting Office, 1977.

Jabber, Paul. "U.S. Interests and Regional Security in the Middle East." *Daedalus* 109 (Fall 1980): 67–80.

Jay, Peter. "Regionalism as Geopolitics." *Foreign Affairs* 57 (America and the World, 1979): 485–514.

Kaiser, Robert G. "U.S.-Soviet Relations: Goodbye to Détente." *Foreign Affairs* 58 (American and the World, 1980): 500–21.

Kalb, Marvin, and Bernard Kalb. "Twenty Days in October." *The New York Times Magazine*, June 23, 1974.

Karnow, Stanley. "Russian Roulette." *New Republic* 116, (October 27, 1973): 12–13.

Keatley, Robert. "Is the Middle East Russia's Vietnam." *Current*, no. 121 (September 1970), pp. 54–56.

Khalilzad, Zalamy. "The Superpowers and the Northern Tier." *International Security* 4 (Winter 1979–80): 6–30.

Kime, Steve F. "The Soviet View of War." *Comparative Strategy* 2 (1980): 205–21.

Kissinger, Henry. *The White House Years.* Boston: Little, Brown, 1979.

________. *Years of Upheaval.* Boston: Little, Brown, 1982.

Kohler, Foy, Leon Goure, and Mose Harvey. *The Soviet Union and the October, 1973 Middle East War.* Miami: University of Miami Press, 1974.

Laquer, Walter. "Detente: What's Left of It." *The New York Times Magazine,* December 16, 1973.
Ledeen, Michael, and William H. Lewis. "Carter and the Fall of the Shah." *The Washington Quarterly* 3 (Spring 1980): 3–41.
Lenczowski, George. "Egypt and the Soviet Exodus." *Current History* 64 (January 1973): 13–16.
Levgold, Robert. "Containment Without Confrontation." *Foreign Policy* no. 40 (Fall 1980), pp. 74-98.
Leiber, Robert. "Energy, Economics, and Security in Alliance Perspective." *International Security* 4 (Spring 1980): 139–63.
Lynch, John B. "The Superpowers' Tug of War Over Yemen." *Military Review* 61 (March 1981): 10–21.
Mansur, Abdul Kasim. "The Military Balance in the Persian Gulf: Who Will Guard the Gulf States from Their Guardians." *Armed Forces Journal International* 118 (November 1980): 44–86.
Marcella, Gabriel. "The Soviet-Cuban Relationship: Symbiotic or Parasitic." Paper Presented for the U.S. Army War College Strategic Studies Institute Foreign Policy Symposium, September 1979, Carlyle Barracks, Penn.
McGuire, Michael. "The Rationale for the Development of Soviet Seapower." *U.S. Naval Institute Proceedings* 106 (May 1980): 155–83.
McGuire, Michael, ed. *Soviet Naval Developments: Capabilities and Context.* New York: Praeger, 1973.
Meir, Golda. "Israel in Search of Lasting Peace." *Foreign Affairs* 51 (April 1973): 447–61.
Middleton, Drew. "Who Lost the Yom Kippur War?" *The Atlantic* 233 (March 1974): 45–66.
Millar, T. B. "Soviet Policies South and East of Suez." *Foreign Affairs* 49 (October 1970): 70–80.
Monroe, Elizabeth, and A. H. Farrar-Hockley. *The Arab-Israeli War, October, 1973.* London: International Institute for Strategic Studies, 1975.
Moore, John Norton, ed. *The Arab-Israeli Conflict.* Princeton: Princeton University Press, 1974.
Morgenthau, Hans J. "We Are Deluding Ourselves in Vietnam." In *Vietnam: Anatomy of a Conflict,* edited by Wesley R. Fishel. Itasca, Ill: Peacock Press, 1968.
Moss, Robert. "On Standing Up to the Russians in Africa." *Policy Review.* no. 5 (Summer 1980), pp. 97–117.
Nitze, Paul J. "Strategy in the Decade of the 1980s." *Foreign Affairs* 59 (Fall 1980): 82–101.
Nixon, Richard M. *The Memoirs of Richard Nixon.* New York: Grosset & Dunlap, 1978.
Noorani, A.G. "Soviet Ambitions in South Asia." *International Security* 4 (Winter 1979–80): 31–59.

O'Ballance, Edgar. "The Fifth Arab-Israeli War—October, 1973." *Army Quarterly and Defence Journal* 104 (July 1976): 308-31.
Odom, William E. "A Dissenting View on the Group Approach to Soviet Politics." *World Politics* 28 (July 1976): 542-67.
_______ . "Whither the Soviet Union." *The Washington Quarterly* 4 (Spring 1981): 30-49.
Park, Chang Jin, "The Influence of Small Powers upon the Superpowers: United States - South Korea Relations as a Case Study, 1950-53." *World Politics* 28 (October 1975): 69-96.
Perlmutter, Amos. "Israel's Fourth War, October, 1973, Political and Military Misperceptions." *Orbis* 19 (Fall 1975): 434-60.
_______ . "The Yemen Strategy." *New Republic* 183, (June 5 and 12, 1980): 16-17.
Pfaltzgraff, Robert L., Jr. "China, Soviet Strategy, and American Policy." *International Security* 5 (Fall 1980): 24-48.
Pike, Douglas. "Communist Vs. Communist in Southeast Asia." *International Security* 4 (Summer 1979): 20-38.
_______ . "The USSR and Vietnam." Paper presented for the U.S. Army War College Strategic Studies Institute Military Policy Symposium, September 1979, Carlyle Barracks Penn.
Pipes, Richard. "Militarism and the Soviet State." *Daedalus* 109 (Fall 1980): 1-12.
Podhertz, Norman. "The Present Danger." *Commentary* 69 (March 1980): 27-39.
Ra'anan, Uri. "The USSR and the Middle East." *Orbis* 17 (Fall 1973): 946-77.
Reagan, Ronald. "Recognizing the Israeli Asset." *The Washington Post*, August 15, 1979.
Remneck, Richard B. "Soviet Policy in the Horn of Africa: The Decision to Intervene." Paper prepared for the U.S. Army War College Strategic Studies Institute Military Policy Symposium, September 1979, Carlyle Barracks, Penn.
Rosecrance, Richard N. "Bipolarity, Multipolarity, and the Future." In *International Politics and Foreign Policy*, edited by James N. Rosenau. New York: The Free Press, 1969.
Rostow, Eugene. *Vital Speeches* 40 (December 1, 1973): 103-6.
Rothenberg, Morris. *The USSR and Africa: New Dimensions of Global Power.* Miami: University of Miami Press, 1980.
Rothstein, Robert L. *Alliances and Small Powers* New York: Columbia University Press, 1968.
Rubinstein, Alvin Z. "Egypt's Foreign Policy." *Current History* 66 (February 1974): 53-56.
_______ . "The Evolution of Soviet Strategy in the Middle East. *Orbis* 24 (Summer 1980): 323-37.

———. *Red Star on the Nile*. Princeton: Princeton Unversity Press, 1977.

———. "The Soviet Union and the Eastern Mediterranean." *Orbis* 23 (Summer 1979): 299–315.

Sabrosky, Alan Ned. "Allies, Clients and Encumbrances." *International Security Review* 5, (Summer 1980): 117–49.

Sadat, Anwar, "From the Memoirs of President as-Sadat." 15-part series published in *October* (in Arabic), October, 31, 1976 through February 5, 1977.

———. *In Search of Identity*. New York: Harper & Row, 1977.

Safire, William. "The Road to Moscow." *New York Times*, October 20, 1979.

Safran, Nadav. "The Soviet Union and Israel, 1947–1969." In *The Soviet Union and the Middle East*, edited by Ivo Lederer and Wayne S. Vucinich. Stanford: The Hoover Institution Press, 1974).

Safran, Nadav. "The War and the Future of the Arab-Israeli Conflict." *Foreign Affairs* 52 (January 1974): 215–36.

Samuels, Michael A., Chester A. Crocker, Roger W. Fontaine, Dimitri K. Simes, and Robert E. Henderson. *Implications of Soviet and Cuban Activities in Africa for U.S. Policy*. Washington: The Center for Strategic and International Studies, 1979.

Sax, Samuel W., and Avigdor Levy. "Arab-Israeli Conflict Four: A Preliminary Assessment." *Naval War College Review* 26, (January-February 1974): 7–16.

Schefield, Victoria. *Bhutto: Trial and Execution*. London: Cassell Ltd., 1979.

Schelling, Thomas, C. *Strategy of Conflict*. Cambridge, Mass.: Harvard University Press, 1960.

Sheehan, Edward R. "Why Sadat Packed Off the Russians." *The New York Times Magazine*, August 6, 1972.

Sheehy, Ann. *The National Composition of the Population of the USSR According to the Census of 1979*. Washington: Radio Liberty/Radio Free Europe, 1980.

Shepherd, George. "Demilitarization Proposals for the Indian Ocean." In *The Indian Ocean in Global Politics*, edited by Larry W. Bowman and Ian Clark, pp. 238–41. Boulder: Westview Press, 1981.

Sicherman, Harvey. "The United States and Israel: The Strategic Divide." *Orbis* 24 (Summer 1980): 381–93.

Simes, Dimitri. "Deterrence and Coercion in Soviet Policy." *International Security* 5 (Winter 1980–81): 80–103.

Singer, Marshall. *Weak States in a World of Powers*. The Free Press, 1972.

Singh, Pushpandar. "JAGUAR Enters Service With India's Air Force." *Armada International* 4 (February 1980): 81–82.

Slominski, Martin J. "The Soviet Military Press and the October War." *Military Review* 54 (May 1974): 39–47.

Smolansky, Oles M. "Soviet Setback in the Middle East." *Current History* 64 (January 1973): 17–20.

Snyder, Glenn H., and Paul Dessing. *Conflict Among Nations.* Princeton: Princeton University Press, 1977.

Sobel, Lester A. ed. *Israel and the Arabs: The October, 1973 War.* New York: Facts on File, 1974.

Spanier, John. *Games Nations Play.* New York: Holt, Rinehart & Winston, 1981.

Sprout, Harold, and Margaret Sprout. "Environmental Factors in the Study of International Politics." In *International Politics and Foreign Policy*, edited by James N. Rosenau, pp. 44–56.

Strategic Survey: 1970. London: The International Institute for Strategic Studies, 1971.

Tucker, Robert W. "America in Decline: The Foreign Policy of 'Maturity.' " *Foreign Affairs* 57 (America and the World, 1979): 449–84.

_______. "American Power and the Persian Gulf." *Commentary* 70 (November 1980): 25–41.

Turbville, Graham H. "Soviet Desert Operations." *Military Review* 54 (June 1974): 40–50.

Turley, William S., and Jeffrey Race. "The Third Indochina War." *Foreign Policy*, 8 (Spring 1980), pp. 92–111.

Vanneman, Peter, and Martin James. "Soviet Intervention in the Horn of Africa." *Policy Review* no. 5 (Summer 1978), pp. 15–37.

Vatikiotis, P.J. "The Soviet Union and Egypt: The Nasser Years." In *The Soviet Union and the Middle East*, edited by Ivo Lederer and Wayne S. Vucinich, Stanford: The Hoover Institution Press, 1974.

Wakebridge, Charles. "The Technological Gap Closes in the Middle East." *National Defense* 59 (May–June 1975): 460–63.

Waltz, Kenneth. "International Structure, National Force, and the Balance of World Power." In *International Politics and Foreign Policy*, edited by James N. Rosenau. New York: The Free Press, 1969.

Wesson, Robert G. "The Soviet Interest in the Middle East." *Current History* 59 (October 1970): 212–19.

Williams, Phil. *Crisis Management.* New York: Wiley, 1976.

Yodfat, Aryeh. *Arab Politics in the Soviet Mirror.* Jerusalem: Israel Universities Press, 1973.

Young, Oran R. *The Politics of Force.* Princeton: Princeton University Press, 1968.

Zagoria, Donald. "Into the Breach: New Soviet Alliances in the Third World." *Foreign Affairs* 57 (Spring 1979): 733–54.

_______. "Soviet Policy and Prospects in the East Asia." *International Security* 5 (Fall 1980): 66–78.

PERIODICALS

Al Ahram, 1969–73.
Baltimore Sun, 1970–80.
Christian Science Monitor, 1972–80.
Congressional Quarterly, 1979–80.
Deadline Data on World Affairs, Soviet Union Foreign Relations, 1959–73.
Deadline Data on World Affairs, United Arab Republic, 1967–73.
Department of State Bulletin, 1970–74.
Economist, 1970–80.
Foreign Broadcast Information Service, Middle East and North Africa, 1967–81.
Foreign Broadcast Information Service, USSR International Affairs, 1967–1981.
Los Angeles Times, 1970–74.
Manchester Guardian, 1970–74.
New York Times, 1955–81.
Observer, 1970.
TASS International Service, 1967–74.
Time, 1973–75–82.
U.S. News and World Report, 1970–74.
Washington Post, 1967–81.

INDEX

ABOUT THE AUTHORS

A professional Army Officer, **Christopher C. Shoemaker** has served in a variety of military assignments in Germany and in the United States. From 1978-79, Shoemaker served on the staff of the Assistant Secretary of Defense for International Security Affairs in the Office of the Near East, Africa, and Southwest Asia. From 1979-82, he was a member of the staff of the National Security Council at the White House, serving under both Presidents Carter and Reagan. In that capacity, he was responsible for defense policy and security issues in the Middle East, Persian Gulf, and Southwest Asia.

Shoemaker received the B.S. from the U.S. Military Academy at West Point, and the M.A. and Ph.D. from the University of Florida.

John Spanier received the B.A. from Harvard University and Ph.D. from Yale University. Since then, he has taught at the University of Florida and has been a Visiting Professor at Haverford College, University of Texas, and Utah State University. Currently, he is a member of the Department of Strategy at Naval War College in Newport, Rhode Island.

Dr. Spanier has lectured at West Point, Naval War College, Maxwell Air Force Base, Foreign Service Institute, and for the U.S. Information Agency in Poland and West Germany.

Among Spanier's former pubications are THE TRUMAN-MACARTHUR CONTROVERSY AND THE KOREAN WAR (Harvard University Press); AMERICAN FOREIGN POLICY SINCE WORLD WAR II, WORLD POLITICS IN AN AGE OF REVOLUTION, GAMES NATIONS PLAY (all published by Holt, Rinehart, and Winston); THE POLITICS OF DISARMAMENT (Praeger); and THE CONGRESS, THE PRESIDENCY AND FOREIGN POLICY (Pergamon). He is also the coauthor of THE DILEMMAS OF AMERICAN FOREIGN POLICY (Holt, Rinehart, and Winston).